8/7/90

✓

UCLA Symposia on Molecular and Cellular Biology, New Series

Series Editor, C. Fred Fox

RECENT TITLES

Volume 108
Acute Lymphoblastic Leukemia, Robert Peter Gale and Dieter Hoelzer, *Editors*

Volume 109
Frontiers of NMR in Molecular Biology, David Live, Ian M. Armitage, and Dinshaw Patel, *Editors*

Volume 110
Protein and Pharmaceutical Engineering, Charles S. Craik, Robert J. Fletterick, C. Robert Matthews, and James A. Wells, *Editors*

Volume 111
Glycobiology, Ernest G. Jaworski and Joseph R. Welply, *Editors*

Volume 112
New Directions in Biological Control: Alternatives for Suppressing Agricultural Pests and Diseases, Ralph R. Baker and Peter E. Dunn, *Editors*

Volume 113
Immunogenicity, Charles A. Janeway, Jr., Jonathan Sprent, and Eli Sercarz, *Editors*

Volume 114
Genetic Mechanisms in Carcinogenesis and Tumor Progression, Curtis Harris and Lance A. Liotta, *Editors*

Volume 115
Growth Regulation of Cancer II, Marc E. Lippman and Robert B. Dickson, *Editors*

Volume 116
Transgenic Models in Medicine and Agriculture, Robert B. Church, *Editor*

Volume 117
Early Embryo Development and Paracrine Relationships, Susan Heyner and Lynn M. Wiley, *Editors*

Volume 118
Cellular and Molecular Biology of Normal and Abnormal Erythroid Membranes, Carl M. Cohen and Jiri Palek, *Editors*

Volume 119
Human Retroviruses, Jerome E. Groopman, Irvin S.Y. Chen, Myron Essex, and Robin A. Weiss, *Editors*

Volume 120
Hematopoiesis, David W. Golde and Steven C. Clark, *Editors*

Volume 121
Defense Molecules, John J. Marchalonis and Carol L. Reinisch, *Editors*

Volume 122
Molecular Evolution, Michael T. Clegg and Stephen J. O'Brien, *Editors*

Volume 123
Molecular Biology of Aging, Caleb E. Finch and Thomas E. Johnson, *Editors*

Volume 124
Papillomaviruses, Peter M. Howley and Thomas R. Broker, *Editors*

Volume 125
Developmental Biology, Eric H. Davidson, Joan V. Ruderman, and James W. Posakony, *Editors*

Volume 126
Biotechnology and Human Genetic Predisposition to Disease, Charles R. Cantor, C. Thomas Caskey, Leroy E. Hood, Daphne Kamely, and Gilbert S. Omenn, *Editors*

Volume 127
Molecular Mechanisms in DNA Replication and Recombination, Charles C. Richardson and I. Robert Lehman, *Editors*

Volume 128
Nucleic Acid Methylation, Gary A. Clawson, Dawn B. Willis, Arthur Weissbach, and Peter A. Jones, *Editors*

Volume 129
Plant Gene Transfer, Christopher J. Lamb and Roger N. Beachy, *Editors*

Volume 130
Parasites: Molecular Biology, Drug and Vaccine Design, Nina M. Agabian and Anthony Cerami, *Editors*

Volume 131
Molecular Biology of the Cardiovascular System, Robert Roberts and Joseph F. Sambrook, *Editors*

Volume 132
Obesity: Towards a Molecular Approach, George A. Bray, Daniel Ricquier, and Bruce M. Spiegelman, *Editors*

Volume 133
Structural and Organizational Aspects of Metabolic Regulation, Paul A. Srere, Mary Ellen Jones, and Christopher K. Mathews, *Editors*

Please contact the publisher for information about previous titles in this series.

UCLA Symposia Board

Frontiers of NMR in Molecular Biology

Frontiers of NMR in Molecular Biology

Proceedings of an Abbott Laboratories–UCLA Symposia Colloquium, Held at Park City, Utah, January 12-19, 1989

Editors

David Live
Department of Chemistry
Emory University
Atlanta, Georgia

Ian M. Armitage
Departments of Molecular Biochemistry, Biophysics, and Diagnostic Radiology
Yale University School of Medicine
New Haven, Connecticut

Dinshaw Patel
Department of Biochemistry
Columbia University
New York, New York

A JOHN WILEY & SONS, INC., PUBLICATION
New York • Chichester • Brisbane • Toronto • Singapore

Address all Inquiries to the Publisher
Alan R. Liss, Inc., 41 East 11th Street, New York, NY 10003

Printed in United States of America

The publication of this volume was facilitated by the authors and editors who submitted the text in a form suitable for direct reproduction without subsequent editing or proofreading by the publisher.

Library of Congress Cataloging-in-Publication Data

UCLA Colloquium, Frontiers of NMR in Molecular Biology (1989 : Park City, Utah)
Frontiers of NMR in molecular biology : proceedings of a UCLA Colloquium held at Park City, Utah, January 12-19, 1989 / editors, David Live, Ian M. Armitage, Dinshaw Patel.
p. cm. -- (UCLA symposia on molecular and cellular biology ; new. ser., v. 109)
"1989 UCLA Colloquium, Frontiers of NMR in Molecular Biology"--Pref.
Includes bibliographical references.
ISBN 0-471-56731-0
1. Nuclear magnetic resonance spectroscopy--Congresses. 2. Molecular biology--Methodology--Congresses. I. Live, David. II. Armitage, Ian M. III. Patel, Dinshaw. IV. University of California, Los Angeles. V. Title. VI. Series.
[DNLM: 1. Molecular Biology--congresses. 2. Nuclear Magnetic Resonance--congresses. W3 U17N new ser. v. 109 / QU 25 U17f 1989]
QP519.9.N83U29 1989
574.8'8'028--dc20
DNLM/DLC
for Library of Congress 89-70452
CIP

Contents

V. DISCUSSION SUMMARIES

Contributors

Celerino Abad-Zapatero, Pharmaceutical Discovery Division, Abbott Laboratories, Abbott Park, IL 60064 **[51]**

Ian M. Armitage, Departments of Molecular Biophysics and Biochemistry and Diagnostic Radiology, Yale University School of Medicine, New Haven, CT 06510 **[xvii, 27]**

R. Boelens, Department of Chemistry, University of Utrecht, 3584 CH Utrecht, The Netherlands **[119]**

Brandan A. Borgias, Departments of Pharmaceutical Chemistry and Radiology, University of California, San Francisco, CA 94143 **[177]**

Aksel A. Bothner-By, Department of Chemistry, Carnegie-Mellon University, Pittsburgh, PA 15213 **[15]**

Michael M. Burns, Rowland Institute for Science, Cambridge, MA 02142 **[167]**

A. Patricia Campbell, Department of Biochemistry and MRC Group of Protein Structure and Function, University of Alberta, Edmonton, Alberta, Canada T6G 2H7 **[99]**

George W. Carter, Pharmaceutical Discovery Division, Abbott Laboratories, Abbott Park, IL 60064 **[75]**

David A. Case, Department of Molecular Biology, Research Institute of Scripps Clinic, La Jolla, CA 92037 **[1]**

E. Charretier, Groupe de Biophysique, Ecole Polytechnique, 91128 Palaiseau, France **[225]**

Kimberly L. Colson, Department of Pharmacology, Yale University School of Medicine, New Haven, CT 06510 **[27]**

A.M. de Vos, Department of Chemistry and Lawrence Berkeley Laboratory, University of California, Berkeley, CA 94720 **[145]**

H. Jane Dyson, Department of Molecular Biology, Research Institute of Scripps Clinic, La Jolla, CA 92037 **[1]**

John W. Erickson, Pharmaceutical Discovery Division, Abbott Laboratories, Abbott Park, IL 60064 **[51]**

Victoria A. Feher, Department of Molecular Biology, Research Institute of Scripps Clinic, La Jolla, CA 92037 **[1]**

The numbers in brackets are the opening page numbers of the contributors' articles.

Juli Feigon, Department of Chemistry and Biochemistry and the Molecular Biology Institute, University of California, Los Angeles, CA 90024 **[249]**

Stephen W. Fesik, Pharmaceutical Discovery Division, Abbott Laboratories, Abbott Park, IL 60064 **[51, 75, 89]**

T.M. Fieser, Department of Molecular Biology, Research Institute of Scripps Clinic, La Jolla, CA 92037 **[63]**

Robert T. Gampe, Jr., Pharmaceutical Discovery Division, Abbott Laboratories, Abbott Park, IL 60064 **[89]**

Dara Gilbert, Department of Chemistry and Biochemistry and the Molecular Biology Institute, University of California, Los Angeles, CA 90024 **[249]**

Jonathan Greer, Pharmaceutical Discovery Division, Abbott Laboratories, Abbott Park, IL 60064 **[75]**

M. Guéron, Groupe de Biophysique, Ecole Polytechnique, 91128 Palaiseau, France **[225]**

Dennis Hare, Hare Research Inc., Woodinville, WA 98072 **[37]**

Edward Hawrot, Department of Pharmacology, Yale University School of Medicine, New Haven, CT 06510 **[27]**

Jeffrey C. Hoch, Rowland Institute for Science, Cambridge, MA 02142 **[167]**

Robert S. Hodges, Department of Biochemsitry and MRC Group of Protein Structure and Function, University of Alberta, Edmonton, Alberta, Canada T6G 2H7 **[99]**

R.A. Houghten, Department of Molecular Biology, Research Institute of Scripps Clinic, La Jolla, CA 92037 **[63]**

S. Hyberts, Biophysics Research Division, University of Michigan, Ann Arbor, MI 48109 **[129]**

Thomas L. James, Departments of Pharmaceutical Chemistry and Radiology, University of California, San Francisco, CA 94143 **[177, 271]**

J. Jancarik, Department of Chemistry and Lawrence Berkeley Laboratory, University of California, Berkeley, CA 94720 **[145]**

R. Kaptein, Department of Chemistry, University of Utrecht, 3584 CH Utrecht, The Netherlands **[119]**

Bo Kim, Department of Chemistry, University of Maryland Baltimore County, Baltimore, MD 21228 **[37]**

S.-H. Kim, Department of Chemistry and Lawrence Berkeley Laboratory, University of California, Berkeley, CA 94720 **[145]**

M. Kochoyan, Groupe de Biophysique, Ecole Polytechnique, 91128 Palaiseau, France **[225]**

R.M.J.N. Lamerichs, Department of Chemistry, University of Utrecht, 3584 CH Utrecht, The Netherlands **[119]**

Richard A. Lerner, Department of Molecular Biology, Research Institute of Scripps Clinic, La Jolla, CA 92037 **[1, 63]**

J.L. Leroy, Groupe de Biophysique, Ecole Polytechnique, 91128 Palaiseau, France **[225]**

David H. Live, Department of Chemistry, Emory University, Atlanta, GA 30322 **[xvii, 259]**

Barbara W. Low, Department of Biochemistry and Molecular Biophysics, College of Physicians and Surgeons, Columbia University, New York, NY 10025 **[15]**

Jay R. Luly, Pharmaceutical Discovery Division, Abbott Laboratories, Abbott Park, IL 60064 **[51]**

Wlodek Mandecki, Corporate Molecular Biology Division, Abbott Laboratories, Abbott Park, IL 60064 **[75]**

Brian J. Marsden, Department of Biochemistry and MRC Group of Protein Structure and Function, University of Alberta, Edmonton, Alberta, Canada T6G 2H7 **[99]**

M.V. Milburn, Department of Chemistry and Lawrence Berkeley Laboratory, University of California, Berkeley, CA 94720 **[145]**

P.K. Mishra, Department of Chemistry, Carnegie-Mellon University, Pittsburgh, PA 15213 **[15]**

K. Miura, Department of Chemistry and Lawrence Berkeley Laboratory, University of California, Berkeley, CA 94720; present address: Faculty of Pharmaceutical Sciences, Hokkaido University, Sapporo, Japan **[145]**

Karl W. Mollison, Pharmaceutical Discovery Division, Abbott Laboratories, Abbott Park, IL 60064 **[75]**

G.T. Montelione, Biophysics Research Division, University of Michigan, Ann Arbor, MI 48109 **[129]**

Peter B. Moore, Departments of Chemistry and Molecular Biophysics and Biochemistry, Yale University, New Haven, CT 06511 **[215]**

David G. Nettesheim, Pharmaceutical Discovery Division, Abbott Laboratories, Abbott Park, IL 60064 **[75]**

N.R. Nirmala, Biophysics Research Division, University of Michigan, Ann Arbor, MI 48109 **[129]**

S. Nishimura, Department of Chemistry and Lawrence Berkeley Laboratory, University of California, Berkeley, CA 94720; present address: Biology Division, National Cancer Center Research Institute, Tokyo, Japan **[145]**

S. Noguchi, Department of Chemistry and Lawrence Berkeley Laboratory, University of California, Berkeley, CA 94720; present address: Biology Division, National Cancer Center Research Institute, Tokyo, Japan **[145]**

E. Ohtsuka, Department of Chemistry and Lawrence Berkeley Laboratory, University of California, Berkeley, CA 94720; present address: Faculty of Pharmaceutical Sciences, Hokkaido University, Sapporo, Japan **[145]**

Edward T. Olejniczak, Pharmaceutical Discovery Division, Abbott Laboratories, Abbott Park, IL 60064 **[75, 89]**

S. J. Opella, Department of Chemistry, University of Pennsylvania, Philadelphia, PA 19104 **[109, 275]**

J.M. Ostresh, Department of Molecular Biology, Research Institute of Scripps Clinic, La Jolla, CA 92037 **[63]**

A. Padilla, Department of Chemistry, University of Utrecht, 3584 CH Utrecht, The Netherlands **[119]**

Dinshaw Patel, Department of Biochemistry, Columbia University, New York, NY 10025 **[xvii]**

Jeffrey G. Pelton, Department of Chemistry, University of California, and Chemical Biodynamics Division, Lawrence Berkeley Laboratory, Berkeley, CA 94720 **[239]**

Ponni Rajagopal, Department of Chemistry and Biochemistry and the Molecular Biology Institute, University of California, Los Angeles, CA 90024 **[249]**

Christina Redfield, Inorganic Chemistry Laboratory, University of Oxford, Oxford, OX1 3QR, England **[155, 167]**

J. Richards, Division of Chemistry and Chemical Engineering, California Institute of Technology, Pasadena, CA 91125 **[109]**

H. Rüterjans, Institut für Biophysikalische Chemie, Johann Wolfgang Goethe-Universität, D-6000 Frankfurt 70, Federal Republic of Germany **[119]**

Harold A. Scheraga, Baker Laboratory of Chemistry, Cornell University, Ithaca, NY 14853-1301 **[189]**

M. Schnarr, Institut de Biologie Moléculaire et Cellulaire du CNRS, Laboratoire de Biophysique, 67084 Strasbourg Cédex, France **[119]**

P. Schrader, Department of Chemistry, University of Pennsylvania, Philadelphia, PA 19104 **[109]**

K. Shon, Department of Chemistry, University of Pennsylvania, Philadelphia, PA 19104 **[109]**

G-Q. Song, Department of Pharmacology, Yale University School of Medicine, New Haven, CT 06510 **[27]**

Terri L. South, Department of Chemistry, University of Maryland Baltimore County, Baltimore, MD 21228 **[37]**

Michael F. Summers, Department of Chemistry, University of Maryland Baltimore County, Baltimore, MD 21228 **[37]**

Brian D. Sykes, Department of Biochemistry and MRC Group of Protein Structure and Function, University of Alberta, Edmonton, Alberta, Canada T6G 2H7 **[99]**

Linda L. Tennant, Department of Molecular Biology, Research Institute of Scripps Clinic, La Jolla, CA 92037 **[1]**

J. Tomich, Division of Chemistry and Chemical Engineering, California Institute of Technology, Pasadena, CA 91125; present address: Division of Medical Genetics, Children's Hospital of Los Angeles, Los Angeles, CA 90027 **[109]**

L. Tong, Department of Chemistry and Lawrence Berkeley Laboratory, University of California, Berkeley, CA 94720 **[145]**

P. Tsang, Department of Molecular Biology, Research Institute of Scripps Clinic, La Jolla, CA 92037 **[63]**

Jennifer Van Eyk, Department of Biochemistry and MRC Group of Protein Structure and Function, University of Alberta, Edmonton, Alberta, Canada T6G 2H7 **[99]**

Herman van Halbeek, Complex Carbohydrate Research Center, University of Georgia, Athens, GA 30613 **[195]**

G. Wagner, Biophysics Research Division, University of Michigan, Ann Arbor, MI 48109 **[129]**

Jonathan P. Waltho, Department of Molecular Biology, Research Institute of Scripps Clinic, La Jolla, CA 92037 **[1]**

David E. Wemmer, Department of Chemistry, University of California, and Chemical Biodynamics Division, Lawrence Berkeley Laboratory, Berkeley, CA 94720 **[239]**

Peter E. Wright, Department of Molecular Biology, Research Institute of Scripps Clinic, La Jolla, CA 92037 **[1, 63]**

Erik R.P. Zuiderweg, Pharmaceutical Discovery Division, Abbott Laboratories, Abbott Park, IL 60064 **[51, 75, 89]**

Preface

The past 10 to 15 years have seen voluminous development of sophisticated NMR spectroscopic techniques for the conformational analysis of biopolymers. The basic NMR strategies have been established from the dizzying array of experimental options illustrated in the accompanying map provided by Aksel Bothner-By from his keynote address. The 1989 UCLA Colloquium, **Frontiers of NMR in Molecular Biology,** addressed the arrival of a significant stage in the application of NMR to the study of problems in structural biology: the transition from emphasis on developing the techniques to their widespread applications in elucidating structures at atomic resolution for moderately large systems. With this effectiveness demonstrated in several systems, and with the purpose of describing the current state of development and exploring the future, it seemed timely to bring together at this conference—and the associated one on protein engineering—investigators from a variety of disciplines relating to the application of NMR to molecular biology.

We are only just beginning to see the results of the confluence of techniques in NMR with (1) those from molecular biology-gene cloning and expression, site-specific mutagenesis, and biosynthetic isotopic labeling, or with (2) those from physical chemistry-molecular dynamics, molecular mechanics, and molecular modeling. The effective integration of the wide-ranging knowledge and facilities needed to assemble the three-dimensional architecture, intermolecular interactions, and dynamics of biomolecules containing thousands of atoms will benefit from increased interactions between groups of scientists such as those assembled for these conferences. The continuing evolution of all these areas, individually and collectively, and the success of this Colloquium, will certainly provide inspiration for further gatherings of this nature.

Contributions to this volume reflect the diversity of presentations and cover analysis of conformation, dynamics, and interactions for a variety of molecules. These include DNA, RNA, carbohydrates, peptides, and proteins.

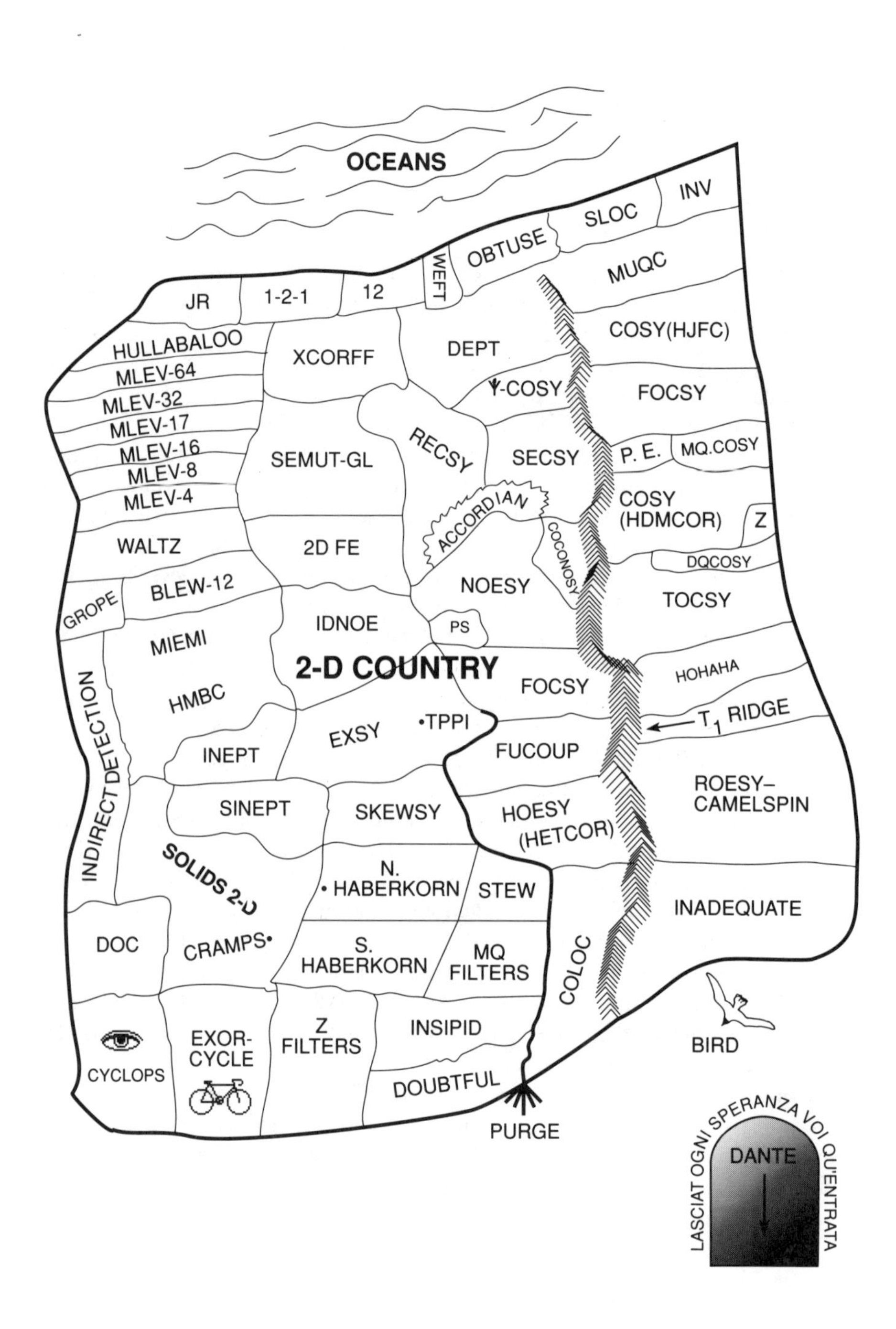
OCEANS
JR
1-2-1
12
WEFT
OBTUSE
SLOC
INV
MUQC
HULLABALOO
MLEV-64
MLEV-32
MLEV-17
MLEV-16
MLEV-8
MLEV-4
XCORFF
DEPT
COSY(HJFC)
Ψ-COSY
FOCSY
RECSY
SEMUT-GL
SECSY
P. E.
MQ.COSY
ACCORDIAN
COSY (HDMCOR)
Z
WALTZ
2D FE
COCONOSY
DQCOSY
NOESY
GROPE
BLEW-12
TOCSY
IDNOE
PS
MIEMI
2-D COUNTRY
HOHAHA
FOCSY
HMBC
T1 RIDGE
•TPPI
EXSY
FUCOUP
INEPT
INDIRECT DETECTION
ROESY– CAMELSPIN
SINEPT
SKEWSY
HOESY (HETCOR)
SOLIDS 2-D
N. • HABERKORN
STEW
INADEQUATE
DOC
CRAMPS•
S. HABERKORN
MQ FILTERS
COLOC
CYCLOPS
EXOR-CYCLE
Z FILTERS
INSIPID
BIRD
DOUBTFUL
PURGE
LASCIAT OGNI SPERANZA VOI QU'ENTRATA
DANTE

We are indebted to those who helped sponsor the Colloquium, particularly the major support from Abbott Laboratories. Additional contributions were received from E.I. du Pont de Nemours & Co., Inc.; Schering Corporation; Smith Kline & French Laboratories; Merck Sharp & Dohme Research Laboratories; Bruker Instruments, Inc.; Merck & Co./Isotopes, and from the Director's Fund established by gifts from Amoco Technology Company, Biotechnology Division; Bristol-Myers Company; Bio-Rad Laboratories, Inc.; DNAX Research Institute; Chiron Corporation; CIBA-Geigy Corporation, Pharmaceuticals Division; Eastman Kodak Life Sciences Research Laboratories; Genetics Institute; Life Technologies, Inc.; New England Biolabs; Pfizer Central Research, Pfizer, Inc.; Pharmacia LKB Biotechnology, Inc.; and Qiagen, Inc. Computer equipment was made available for the electronic poster session by Abbott Laboratories; GE NMR Instruments; Biodesign; Hare Research; New Methods Research, Inc.; Sun Microsystems, Inc.; and Varian Associates. All of these and the exceptional assistance of the UCLA Symposia staff were invaluable in making the meeting a great success.

David Live
Ian M. Armitage
Dinshaw Patel

Frontiers of NMR in Molecular Biology, pages 1-13

FOLDING OF PEPTIDE FRAGMENTS OF PROTEINS IN AQUEOUS SOLUTION[1]

Peter E. Wright, H. Jane Dyson,
Victoria A. Feher, Linda L. Tennant, Jonathan P. Waltho,
Richard A. Lerner and David A. Case

Department of Molecular Biology
Research Institute of Scripps Clinic
10666 North Torrey Pines Road
La Jolla, California 92037

ABSTRACT Recent developments in two-dimensional NMR spectroscopy have provided very sensitive methods for the detection of folded structures in peptide fragments of proteins in water solution. Several types of structure have been observed, including β-turns, nascent helix, extended helical conformations and hydrophobic clusters. The folded conformations are mostly relatively unstable and are in rapid dynamic equilibrium with unfolded states. Folded peptide structures can be identified by the presence of medium- or long-range NOEs, unusual $^3J_{HN\alpha}$ coupling constants and amide proton exchange measurements or temperature coefficients. In some cases, the structures can also be detected by circular dichroism spectroscopy. Molecular dynamics simulations can play an important role in assessing the nature of the interactions which stabilize the folded conformations. The observations that peptide fragments of proteins adopt secondary structures in water solution has profound implications for initiation of protein folding, for the mechanism of induction of protein-reactive antipeptide antibodies and for T cell recognition.

[1]This work was supported by grants GM38794 and CA27489 from the National Institutes of Health.

INTRODUCTION

There is currently intense interest in the conformational properties of linear peptide fragments of proteins in aqueous solution. This arises on one hand from efforts to understand the mechanisms of induction of protein-reactive antipeptide antibodies and on the other hand from a resurgence of interest in the protein folding problem. Early attempts to find folded structures in peptide fragments of proteins were generally unsuccessful (1-3) and led to a general belief that short linear peptides do not contain secondary structure in water solution. For many years, the only exception appeared to be the C-peptide (residues 1-13) of ribonuclease A (4,5). Recently, however, we have been successful in using a combination of antipeptide antibody technology and two-dimensional NMR techniques both to identify peptides which have a high propensity to adopt folded structures in aqueous solution and to obtain direct experimental information on the folded conformations (6,7,8). The realization that many immunogenic peptides exhibit conformational preferences for secondary structure in aqueous solution has important implications for induction of protein-reactive antipeptide antibodies (9) and for initiation of protein folding (10).

Our NMR experiments have provided unequivocal evidence for the formation of reverse turns, helical conformations and hydrophobic clusters in short linear peptides in aqueous solution. The experimental evidence for these structures has been described in detail elsewhere (6-8,11-13). In the present paper, we will address more general questions concerning interpretation of NMR data for flexible linear peptides. In particular, we focus on the reverse turn, to illustrate how a combination of multiple peptide synthesis, NMR spectroscopy, circular dichroism spectroscopy and molecular dynamics simulations can be used to identify preferred conformations of a peptide in aqueous solution and to assess the subtle factors which stabilize those conformations.

METHODS

The NMR experiments used to identify peptide structures have been described in detail elsewhere (7,8). Briefly, resonance assignments are made using standard sequential assignment strategies and several NMR parameters can be used to provide information about peptide conformations. Most importantly, direct information on interproton distances can be inferred from the nuclear Overhauser effect (NOE). Depending on the size of the peptide, NOE data are obtained from NOESY spectra (14) or from rotating frame NOESY (ROESY) spectra (15).

Molecular dynamics simulations of peptides in water were carried out with an extensively modified version of the program AMBER 3.0 (16) using the "OPLS" potential functions developed by Jorgensen and co-workers (17) and the TIP3P model for water (18). The peptides Ac-Ala-Pro-Gly-X-NHMe were considered, where X=Asp (APGD), Ala (APGA) or Asn (APGN). These were initially built in an idealized Type II turn configuration, with (ϕ,ψ) angles of (-60,120) for proline and (90,0) for glycine. These were inserted into a box of pre-equilibrated water molecules such that there was at least 9 Å from any peptide atom to the nearest wall. This led to a system with 419 water molecules for the APGD peptide, 427 for APGA and 452 for APGN. Periodic boundary conditions were applied; the computations were carried out on a Cray XMP with an integration time step of 0.5 fsec. Each picosecond of simulation took about 8 min. of cpu time.

During the initial equilibration period, the hydrogen bond between the carbonyl oxygen of residue 1 and the amide hydrogen of residue 4 was kept near 1.9 Å by a harmonic constraint on the O-H distance. The equilibration took place over 40 psec; details will be given elsewhere. The restraint on the 1-4 "β-turn" hydrogen bond was removed for the "data-collection" runs. For APGA and APGN, additional starting points were obtained by continuing the equilibration (with the O-H restraint present) for additional 10 psec intervals, in order to obtain new phase space points from which to start data collection.

RESULTS AND DISCUSSION

Identification of Peptide Structures by NMR Spectroscopy.

Small linear peptides do not adopt unique conformations in aqueous solution, but sample a number of conformational states. NMR parameters such as the chemical shift and coupling constant are thus a population-weighted average over all conformers and the NOESY spectrum contains cross peaks representative of all conformations that have sufficiently high populations. Fortunately, however, only a relatively limited range of backbone ϕ and ψ dihedral angles is heavily populated and it is thus possible to interpret, at a qualitative level at least, patterns of NOEs involving backbone protons. A general conformational energy diagram (Figure 1) provides an important guide to the interpretation of NOE data for linear peptides. A given peptide fluctuates over an ensemble of conformations with a range of ϕ and ψ angles lying within the minima of Figure 1. Different patterns of NOE connectivities are observed for the peptide

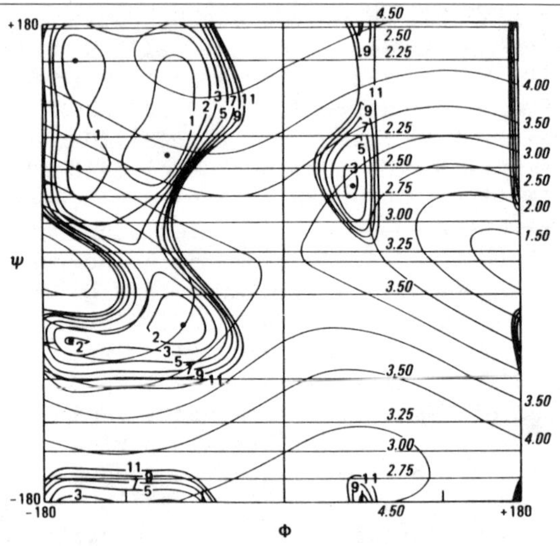

FIGURE 1. Conformational energy diagram for the alanine dipeptide (adapted from (25)). Energy contours are drawn at intervals of 1 kcal.mol^{-1}. The potential energy minima for β, α_R and α_L are labelled. The dependence of the sequential $d_{NN}(i,i+1)$ distance (in Å) on the ϕ and ψ dihedral angles (26) is shown as a set of dashed contours. The $d_{\alpha N}(i,i+1)$ distance depends only on w for trans peptide bonds (26) and is represented as a series of contours parallel to the ϕ axis.

in each of these minima in ϕ,ψ space. Strong $d_{\alpha N}(i,i+1)$ NOE connectivities in the absence of $d_{NN}(i,i+1)$ NOEs show that the backbone dihedral angles are predominantly in the β (extended chain) region of ϕ,ψ conformational space. The observation of $d_{NN}(i,i+1)$ NOE connectivities indicates that the conformational ensemble of the peptide includes local structures with dihedral angles in the α_R or α_L minima. For some peptides, both $d_{\alpha N}(i,i+1)$ and $d_{NN}(i,i+1)$ NOEs are observed between given pairs of residues. When both types of sequential NOE connectivity are observed at a single site in a peptide, this is often an indication of a highly preferred local structure such as a type II β-turn for which the ϕ,ψ angles for residue 3 lie in the α_L region. When extended sequences of $d_{\alpha N}(i,i+1)$ and $d_{NN}(i,i+1)$ NOE connectivities overlap, this is a good indication of conformational averaging, i.e., the peptide samples both the α_R and β regions of ϕ,ψ space.

Observation of sequential $d_{\alpha N}(i,i+1)$ or $d_{NN}(i,i+1)$ NOE connectivities alone does not reliably indicate that *folded* peptide structures are present. Rather, the sequential NOE connectivities provide information about the *local* dihedral angle populations at each residue, i.e., they provide information about the relative populations of the β and α_R minima for each residue in the peptide. Thus, the observation of an extended sequence of $d_{NN}(i,i+1)$ NOE connectivities does not constitute sufficient evidence that a peptide adopts helical conformations in solution; it merely indicates that there is a significant (i.e., detectable) population of ϕ and ψ angles in the α_R region. Reliable identification of folded structures requires additional information, either from NMR spectroscopy or from other forms of spectroscopy such as circular dichroism. The best diagnostics of folded structures are medium-range or long-range NOEs, which immediately identify a significant population of non-random folded structures in the conformational ensemble. These diagnostic NOE connectivities are summarized in Table 1. Additional information on folded structures may come from unusual $^3J_{NH\alpha}$ coupling constants or from evidence that particular protons are involved in hydrogen bonding interactions, as shown by lowered amide proton temperature coefficients or reduced amide proton exchange rates.

TABLE 1
DIAGNOSTIC NOE CONNECTIVITIES IN LINEAR PEPTIDES

Structure	NOE Observed	Interproton Distance
Unfolded states:		
β-region	$d_{\alpha N}(i,i+1)$	2.2 - 2.7 Å
α_R-region	$d_{NN}(i,i+1)$	1.9 - 3.0 Å
Folded structures:		
Helix	$d_{NN}(i,i+1)$	2.6 - 2.8 Å[a]
	$d_{\alpha N}(i,i+3)$	3.3 - 3.4 Å[a]
	$d_{\alpha\beta}(i,i+3)$	2.5 - 5.1 Å[a]
Reverse turn	$d_{NN}(3,4)$	2.4 Å[b]
	$d_{\alpha N}(2,4)$	3.3 - 3.5 Å[b]

[a]Distances in regular α- or 3_{10}-helices (23).
[b]Distances in regular Type I or Type II β-turns (7,23).

Evidence for Reverse Turn Formation

As an example, we consider the reverse turn. The relevant NOE connectivities for the peptide YPGDV are summarized in Figure 2. As for all peptides studied by us, sequential $d_{\alpha N}(i,i+1)$ NOEs (or the equivalent $d_{\alpha\delta}$ NOEs when residue i+1 is Pro) are observed for all residues. These NOEs are relatively uninformative about structured forms since they occur both in folded conformations and in the unfolded (extended chain) state. The observation of a $d_{NN}(i,i+1)$ NOE connectivity between Gly 3 and Asp 4, *together with* a $d_{\alpha N}(i,i+2)$ NOE between Pro 2 and Asp 4 is consistent with formation of a reverse turn in the *trans* isomeric form. Note that these NOEs are not observed for the *cis* isomer which acts as an internal unfolded control. Further NMR evidence for formation of a reverse turn comes

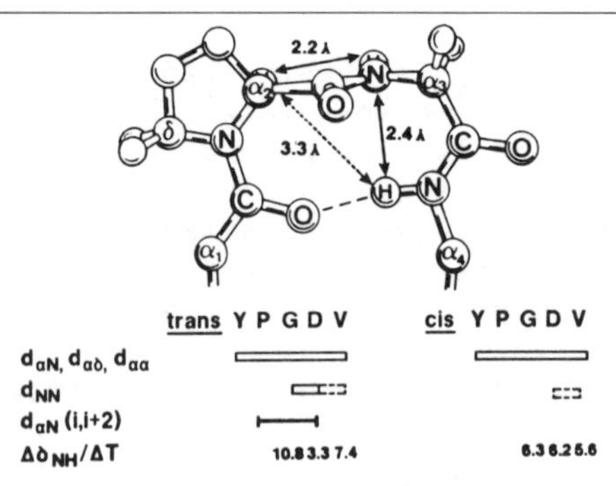

FIGURE 2. Schematic representation of a Type II β-turn in the sequence YPGD, showing the expected short interproton distances; summary of the observed NOEs and temperature coefficients for YPGDV.

from amide proton resonance temperature coefficients, that of the Asp 4 NH being substantially lowered in the folded *trans* form of the peptide but not in the unfolded *cis* isomer (Figure 2). The most likely explanation for this low temperature coefficient is formation of a 1-4 hydrogen bond between the Asp 4 amide proton and the carbonyl of Tyr 1. Thus NMR provides several pieces of evidence that *trans*-YPGDV has a strong conformational preference for a reverse turn in water solution. This turn appears to be closest to a Type II turn since the $d_{\delta N}$(Pro,Gly) NOE connectivity which would be expected for a Type I turn is completely absent from the NOESY spectra of this and related turn-containing peptides (7). Circular dichroism measurements confirm that YPGDV adopts a significant population of β-turn conformations in water solution, as indicated by positive ellipticity with a maximum near 206 nm (7).

The observation of significantly populated reverse turn conformations in such small peptides in aqueous solution is surprising. An NMR study of the population of reverse turn conformations in a series of peptides where positions 1, 3 and 4 were systematically varied has demonstrated a marked dependence of turn stability on amino acid sequence (7). Substitutions at position 3 and 4 can enhance or abolish the reverse turn population in the *trans* peptide isomers. The residue at position 3 of the turn is the primary determinant of its stability. Hydrophilic side chains at position 4 favor turn formation while hydrophobic side chains destabilize the turn. Substitutions at position 1 also have small effects on the turn population. Of the peptides studied, APGDV shows the highest propensity for turn formation in aqueous solution and other amino acid residues which favor helix formation in proteins (19) also appear to favor turn formation when present at position 1. The presence of a deprotonated Asp side chain at position 4 of the turn appears to confer considerable stability, as measured by the temperature coefficient of the hydrogen-bonded NH proton of the residue at position 4 in the *trans* isomer. This is confirmed by circular dichroism measurements on APGDV (Figure 3); the positive maximum near 208 nm characteristic of the β-turn conformation is greatly diminished at low pH.

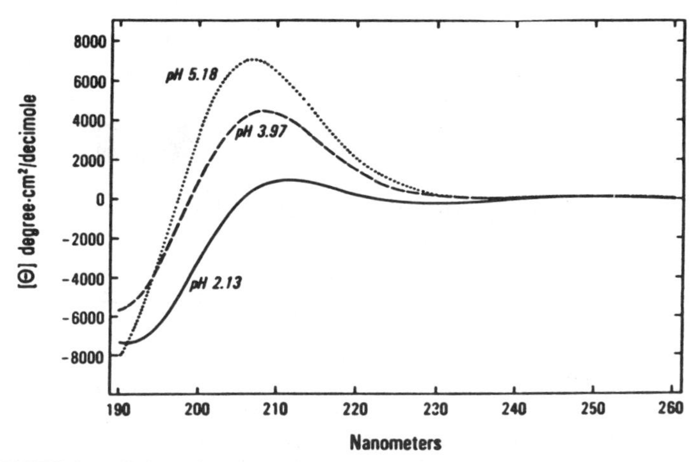

FIGURE 3. pH-dependence of the circular dichroism spectrum of APGDV. The positive ellipticity near 208 nm is characteristic of the β-turn conformation.

Molecular Dynamics Simulations on β-Turn Peptides

Molecular dynamics simulations were carried out to establish that the reverse turn conformations proposed on the basis of the NMR data are energetically feasible and to provide some insight into the molecular interactions which may be responsible for stabilizing the reverse turn.

APGD. The hydrogen bond distances for the APGD simulation are illustrated in Figure 4. From these and from the backbone angles (not shown) it is clear that the basic type-II turn configuration is stable for the first 500 psec, and that no conformational transitions take place during this time. Average backbone angles (ϕ,ψ) for this period are (-60,+110) for proline and (+110,-20) for glycine. The average 1-4 hydrogen bond length is 2.0 Å, but it rises above 2.5 Å 52 times during the first 500 psec, or an average of once every 10 psec.

This local turn is also stabilized by attractive interactions between the carbonyl of residue 1 and the amide proton of residue 5, i.e., a bifurcated hydrogen bond is formed; this pattern is seen in the other sequences as well. There is no direct evidence from the present NMR experiments for a bifurcated hydrogen bond, but these results suggest

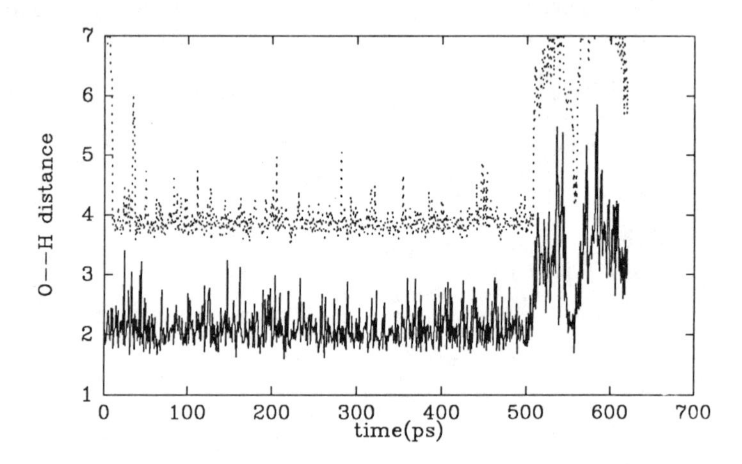

FIGURE 4. Molecular dynamics simulation for Ac-APGD-$NHCH_3$. Time-dependence of the distances between the carbonyl oxygen atom of Ala 1 and the amide proton of Asp 4 (solid line) and between the Ala 1 carbonyl oxygen and the amide hydrogen of the N-methyl "blocking group" (dotted line, shifted up by 2 Å), which would correspond to residue 5 in a longer peptide.

further experiments to test the role of such hydrogen bonding interactions in stabilizing the reverse turn. The APGD turn is not stabilized by any interactions between the charged Asp side chain and the N-terminus of the peptide; instead this side chain is strongly solvated and points away from the rest of the peptide.

After about 510 psec of simulation, a major conformational change takes place in the peptide, as seen by a rapid rise in the O-H distances (Figure 4) and the backbone torsional angles. The system briefly returns to a configuration close to that of type I turn (near 550 psec), but then changes again toward an extended chain with no 1-4 or 1-5 internal hydrogen bonds.

AGPA and APGN. Figure 5 shows results for these sequences with a protocol identical to that used for APGD. In general, the turns in APGN and APGA persisted for shorter times than was the case for APGD. Although of limited statistical significance at this stage, these results are consistent with the observation that peptides with deprotonated Asp in position 4 have a higher turn propensity than do peptides with either Ala or Asn at this position.

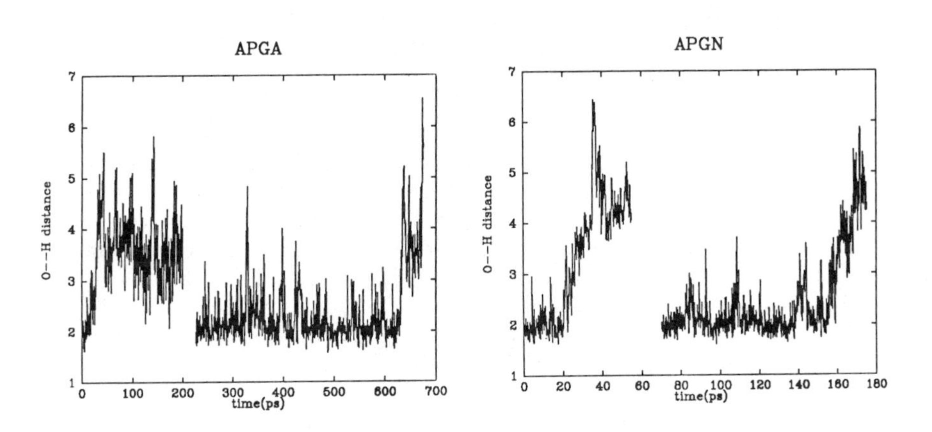

FIGURE 5. (left) Molecular dynamics simulation for Ac-APGA-$NHCH_3$. Time course of the distance between the carbonyl oxygen of Ala 1 and the amide proton of Ala 4. (right) Molecular dynamics simulation for Ac-APGN-$NHCH_3$. Time course of the distance between the carbonyl oxygen of Ala 1 and the amide proton of Asn 4. For each of these peptides, simulations from two independent starting points are shown.

The results of these few simulations cannot be used to provide an estimate of the stability of the turn forms of these peptides. For a straightforward simulation, one would need many "transitions" between folded and unfolded forms in order to determine relative stabilities. Alternatively, one can use biased sampling procedures to map out free energy profiles (20); such methods have been applied to dipeptides (21) and, more recently, to turn and helical peptide structures (22). However, unbiased simulations of the type discussed here are particularly useful for the discovery of potentially stable structures, such as those with bifurcated hydrogen bonds, that might be difficult to predict in advance. Further, a rough idea of the time scale for various motions can be found in the present simulations: a typical transition between a turn and an extended configuration took 20-40 psec in these simulations. This is considerably longer than the transition time scales (< 5 psec) seen in simulations of tyrosine rotations in protein interiors (23) or diffusion of gases through proteins (24), suggesting that conformational transitions even in very short peptides are rather complex events. A combination of constrained and unconstrained simulations seems likely in the near future to provide real theoretical insight into the conformational preferences of short peptides in solution.

Evidence for Helical Conformations

The observation of an extended sequence of $d_{NN}(i,i+1)$ NOE connectivities is not of itself sufficient evidence for the presence of helix. Such NOE connectivities can arise in unfolded peptides which have a substantial population of ϕ and ψ angles in the α_R region of ϕ,ψ conformational space (Figure 1). For reliable identification of helical structures by NMR, observation of medium-range $d_{\alpha\beta}(i,i+3)$ and $d_{\alpha N}(i,i+3)$ NOE connectivities, in addition to the sequential $d_{NN}(i,i+1)$ NOEs, is mandatory. For example, a peptide corresponding to residues 132 to 153 of sperm whale myoglobin populates helical conformations between residues 133 and 149. In favorable cases circular dichroism spectroscopy can provide confirmatory evidence for a multiple turn helix. However, the absence of a characteristic helical CD spectrum, with minima at 222 and 208 nm, does not eliminate helical conformations from consideration. For example, a short helix which is in rapid equilibrium with unfolded states and is located in a much longer peptide may be essentially undetectable by CD methods but can still give rise to characteristic helical NOE connectivities. This is in fact the case for the myoglobin peptide. Further, the characteristic minimum at 222 nm indicative of helix can be masked by ellipticity from the aromatic amino acids. This is illustrated by substitution of Tyr 146 in the myoglobin peptide by lysine. When this substitution is made, the CD spectrum then shows the characteristic minimum at 222 nm indicative of helix although no discernable changes in the population of helix are evident from NMR experiments.

In summary, NMR has emerged as a very powerful method for elucidating the conformational preferences of small linear peptides in aqueous solution. The method is suffi ciently sensitive that it is possible to detect low populations of folded conformations or identify highly localized structures in large peptides that are all but invisible using more conventional spectroscopic techniques such as circular dichroism. However, in favorable cases CD spectra can provide important conformation of the secondary structures deduced from NMR data. The combination of antipeptide antibody technology with two-dimensional NMR methods provides a novel approach for identification and structural characterization of peptide fragments of proteins that have a marked preference for folded structures in aqueous solution. These techniques are complemented by synthetic methods which can produce large numbers of peptides with specific changes in the amino acid sequence. Studies of peptide conformations in aqueous solution can be expected to provide important new insights into the earliest events of protein folding and may eventually reveal the fundamental amino acid sequence code for folding.

REFERENCES

1. Epand RM, Scheraga HA (1968). The influence of long-range interactions on the structure of myoglobin. Biochemistry 7:2864.
2. Taniuchi H, Anfinsen CB (1969). An experimental approach to the study of the folding of staphylococcal nuclease. J Biol Chem 244:3864.
3. Hermans J, Puett D (1971). Relative effects of primary and tertiary structure on helix formation in myoglobin and α-lactalbumin. Biopolymers 10:895.
4. Brown JE, Klee WA (1971). Helix-coil transition of the isolated amino terminus of ribonuclease. Biochemistry 10:470.

5. Shoemaker KR, Kim PS, Brems DN, Marqusee S, York EJ, Chaikin IM, Stewart JM, Baldwin RL (1985). Nature of the charged-group effect on the stability of the C-peptide helix. Proc Natl Acad Sci USA 82:2349.
6. Dyson HJ, Cross KJ, Houghten RA, Wilson IA, Wright PE, Lerner RA (1985). The immunodominant site of a synthetic immunogen has a conformational preference in water for a type-II reverse turn. Nature 318:480.
7. Dyson HJ, Rance M, Houghten RA, Lerner RA, Wright PE (1988). Folding of immunogenic peptide fragments of proteins in water solution. 1. Sequence requirements for the formation of a reverse turn. J Mol Biol 201:161.
8. Dyson HJ, Rance M, Houghten RA, Wright PE, Lerner RA (1988). Folding of immunogenic peptide fragments of proteins in water solution. 2. The nascent helix. J Mol Biol 201:201.
9. Dyson HJ, Lerner RA, Wright PE (1988). The physical basis for induction of protein-reactive antipeptide antibodies. Annu Rev Biophys Biophys Chem 17:305.
10. Wright PE, Dyson HJ, Lerner RA (1988). Conformation of peptide fragments of proteins in aqueous solution: Implications for initiation of protein folding. Biochemistry 27:7167.
11. Waltho JP, Lerner RA, Wright PE. Conformation of a T cell stimulating peptide in aqueous solution. Submitted for publication.
12. Osterhout JJ, Baldwin RL, York EJ, Stewart JM, Dyson HJ, Wright PE. ^{1}H NMR studies of the solution conformations of an analogue of the C-peptide of ribonuclease A. Submitted for publication.
13. Sayre J, Dyson HJ, Lerner RA, Wright PE, unpublished results.
14. Jeener J, Meier BH, Bachmann P, Ernst RR (1979). Investigation of exchange processes by two-dimensional NMR spectroscopy. J Chem Phys 71:4546.
15. Bothner-By AA, Stephens RL, Lee J, Warren CD, Jeanloz RW (1984). Structure determination of a tetrasaccharide: Transient nuclear Overhauser effects in the rotating frame. J Am Chem Soc 106:811.
16. Singh UC, Caldwell J, Weiner P, Kollman PA (1986). "Amber 3.0," University of California, San Francisco; modifications to the code were made by J Kottalam and DA Case.
17. Jorgensen WL, Tirado-Rives J (1988). The OPLS potential functions for proteins. Energy minimizations for

crystals of cyclic peptides and crambin. J Am Chem Soc 110:1657.
18. Jorgensen WL, Chandrasekhar J, Madura JD, Impey RW, Klein ML (1983). Comparison of simple potential functions for stimulating liquid water. J Chem Phys 79:926.
19. Chou PY, Fasman GD (1978). Prediction of the secondary structure of proteins from their amino acid sequence. Adv Enzymol 47:45.
20. Mezei M, Beveridge DL (1986). Free energy simulations. Ann NY Acad Sci 482:1.
21. Beveridge DL, Ravishankar G, Mezei M, Gedulin B (1985). Solvent effects on conformational stability in the Ala dipeptide: Full free energy simulations. In Sarma R (ed): "Biomolecular Stereodynamics III," Albany, New York: Adenine Press, p 237. Anderson AG, Hermans J (1988). Microfolding: Conformational probability map for the alanine dipeptide in water from molecular dynamics simulations. Proteins: Str Func Gen 3:262.
22. Hermans J, personal communication.
23. Northrup SH, Pear MR, Lee CY, McCammon JA, Karplus M (1982). Dynamical theory of activated processes in globular proteins. Proc Natl Sci USA 79:4035. McCammon JA, Lee CY, Northrup SH (1983). Sidechain rotational isomerization in proteins: A mechanism involving gating and transient packing defects. J Am Chem Soc 105:2232.
24. Kottalam J, Case DA (1988). Dynamics of ligand escape from the heme pocket of myoglobin. J Am Chem Soc 110:7690.
25. Ramachandran GN, Venkatachalam CM, Krimm S (1966). Stereochemical criteria for polypeptide and protein chain conformations. III Helical and H-bonded polypeptide chains. Biophys J 6:849.
26. Billeter M, Braun W, Wüthrich K (1982). Sequential resonance assignments in protein ^{1}H nuclear magnetic resonance spectra. Computation of sterically allowed proton-proton distances in single crystal protein conformations. J Mol Biol 155:321.

Frontiers of NMR in Molecular Biology, pages 15-25

INTERACTION OF A SNAKE VENOM NEUROTOXIN AND A SEQUENCE FROM THE ACETYLCHOLINE RECEPTOR α-SUBUNIT[1]

Aksel A. Bothner-By, P. K. Mishra and Barbara W. Low[2]

Department of Chemistry, Carnegie Mellon University
Pittsburgh, PA 15213

ABSTRACT: From a study of the stereochemistry of the reactive site of protein α-toxins considered as a probe, one segment of the α-subunit of the acetylcholine receptor (residues 177-193) has been recognized by Low and Corfield (1) as the prime complementary binding domain. Intrinsic fluoresence studies (2) showed very tight binding ($K_b \cong 5 \times 10^{-8}$M) of the 179-191 region of calf peptide (KESRGWKHWVFYA). The dominant fluorophor, W187, is heavily quenched on binding, suggesting a marked change in environment. We have applied 1-D and 2-D proton spectroscopies (COSY, NOESY) to the study of this peptide, α-cobratoxin, and their mixtures. The results on the toxin agree well with those of Kondakov, et al (3), and with X-ray structure determination (4). The peptide is small and appears rather flexible in solution, though anomalous shifts indicate formation of a hydrophobic region. The suggested structure of the complex is supported by the NMR results.

---

[1]NMR spectra recorded at the NMR Facility for Biomedical Studies, supported by National Institute of Health Grant RR00292. This work was in part supported by NIH Grant NS22719 (BWL).

[2]Department of Biochemistry and Molecular Biophysics, College of Physicians and Surgeons, Columbia University.

INTRODUCTION

Venoms from a variety of land and sea snakes contain neurotoxins which are structurally closely related (5), and which bind to the acetylcholine receptor (AcChR) of the post-synaptic membrane producing neuromuscular block (6,7). The acetylcholine receptor consists of five subunits, with the scheme $\alpha_2\beta\gamma\delta$(8). It has been shown that the prime and competitive binding site is on the α-subunit (9).

Low and Corfield (1) have examined the amino-acid sequences of the subunits of AcChRs from a variety of species, and have identified a unique sequence, V177-C193, which contains side-chains complementary to those in the toxins and one or more tryptophan residues positioned to bind in the cleft of the α-toxins. That this segment contains a sequence which will bind tightly to both short and long chain toxins has been demonstrated by observation of fluorescence quenching of W187 in the calf peptide sequence K179-A191 upon addition of either short-chain (Erabutoxin) or long-chain (Cobratoxin) neurotoxins. In this paper we report the initial results of 1-D and 2-D NMR spectroscopic studies of cobratoxin, the calf peptide, and of mixtures of the two.

COBRATOXIN

A view of cobratoxin is given in Fig. 1, based on the coordinates given by Walkinshaw, et al (4). The main structural features include a 3-strand anti-parallel pleated sheet, consisting of segments C20-W25, R36-C41, and D53-C57, forming the "floor" of the cleft. A "rail" C45-T50 forms a wall enclosing the cleft. It is suggested (1) that residues R33, D27 and D38 form salt bridges respectively with E180, R182 and K185 of the peptide, and that either W184 or W187, with a preference for the latter, is bound in the cleft.

The NMR spectrum at 500 MHz of cobratoxin at pH 3.3, 50° C, has been essentially completely assigned by Kondakov, et al, (3) using a combination of NOESY and COSY techniques. This group has also studied the pH dependence of the spectra, and established that cobratoxin exists in two conformations, A and B, stable at low and high pH, respectively (10). The change is triggered by deprotonation of H18, which then folds into the structure.

A 6.0-8.0 ppm segment of the 1-D proton spectrum of α-cobratoxin in D_2O at pH 6.3, 23° C is shown in Fig. 2, and the corresponding COSY spectrum is shown in Fig. 3.

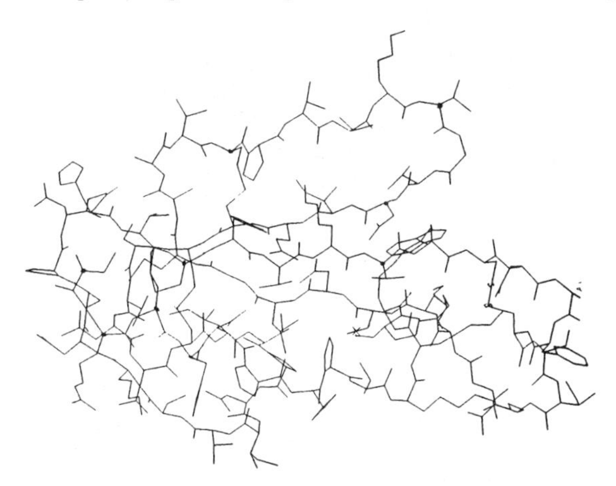

FIGURE 1. View of Cobratoxin according to coordinates of Walkinshaw, et al (3).

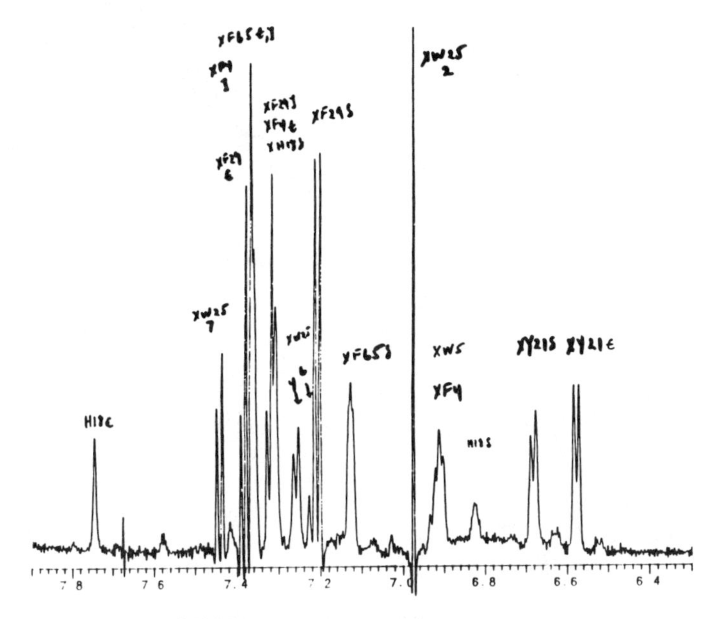

FIGURE 2. Aromatic region, 620 MHz spectrum of cobratoxin in D_2O, pH 6.3, 23^{0}C.

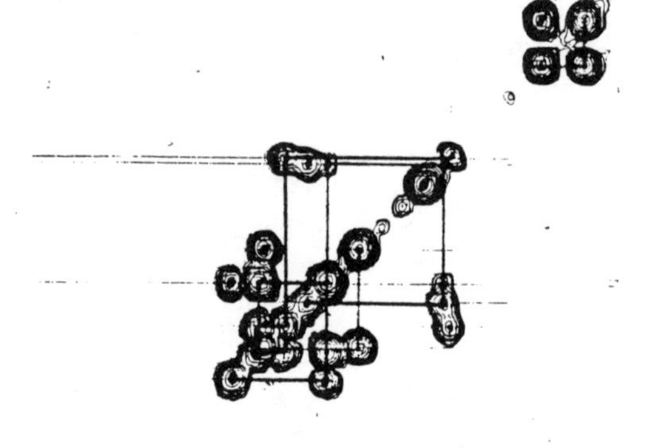

FIGURE 3. COSY spectrum, 620 MHz, aromatic region cobratoxin in D_2O, pH 6.3, 23°C.

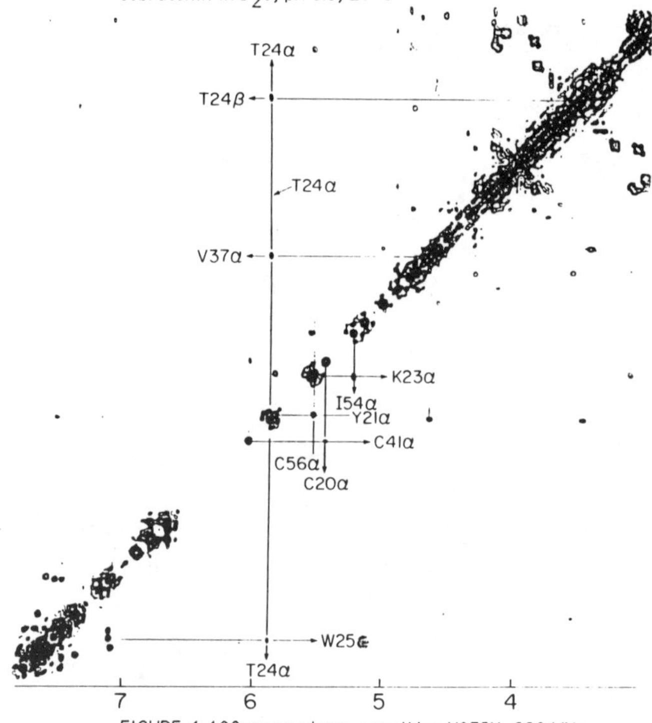

FIGURE 4. 100 msec. phase-sensitive NOESY, 620 MHz, low-field region, cobratoxin in D_2O, pH 6.3, 23°C.

Assignment of the aromatic resonances of the unique Y21 and W25 residues is straightforward. The remaining F residues (4, 29, 65) are assigned following Kondakov, et al (3), although the F65 ε and ζ protons appear to be superposed at 7.36 ppm. (acetate peak at 1.92 ppm).

A 2.0-8.0 ppm segment of the 100 msec. NOESY spectrum of the same sample is displayed in Fig. 4. Confirming the structure already established by X-ray and NMR studies are the NOE effect noted between T24α and W25(4), and the $d_{\alpha\alpha}$ NOES characteristic of the pleated sheet Y21α-C56α, C41α-C20α, T24α,-V32α, and K23α,-I54α.

CALF PEPTIDE

The proton resonance spectrum in the aromatic region is shown in Fig. 5, with assignments indicated. Spectra were obtained at pH 3.4 and at pH 7.0. The assignments were made by standard technques, including the NH-αCH correlations by COSY, titration shifts, and NOE effect measurements. A complete listing of assignments is given

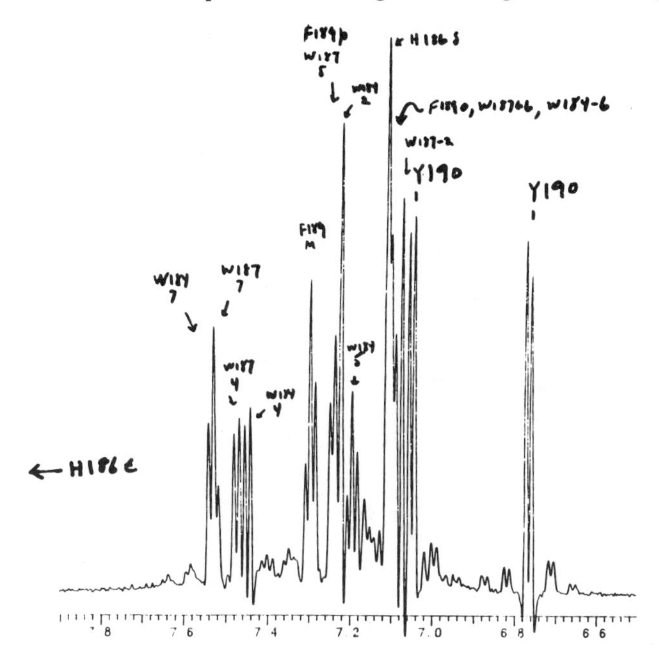

FIGURE 5. Aromatic region, 620 MHz spectrum of calf peptide in D_2O, pH 3.4, 23°C .

in Table I. The shifts observed for the peptide residues are very similar to those observed as average (11), except that the K185 and V188 side chain resonances are shifted upfield by 0.2–0.4 ppm. This would suggest an extended form of the chain in this region with hydrophobic interaction between the side chains of K185, W187 and F189, as well as between H186, V188, and Y190.

TABLE I.

CALF PEPTIDE ASSIGNMENTS

	NH	α	β	γ	δ	ε	ζ	η
K179	--	4.05	~1.89	1.45 1.46	~1.69	2.99		
E180	8.84	4.43	1.95 2.04	2.36				
S181	8.57	4.48	3.82 3.86					
R182	8.52	4.30	1.65 1.80	1.57 1.60	3.13			
G183	8.43	3.90 3.94						
W184	7.97	4.60	3.25 3.27		7.21	7.45	7.19 7.53	7.05
K185	8.10	4.04	~1.48	1.01 1.04	~1.48	2.79		
H186	8.20	4.51	3.02 3.15		7.10	8.47		
W187	8.23	4.59	3.17 3.19		7.07	7.44	7.24 7.52	7.06
V188	7.79	3.96	1.78	0.66 0.75				
F189	8.02	4.40	2.81 2.89		7.07	7.28	7.24	
Y190	7.91	4.46	2.79 2.96		7.05	6.77		
A191	7.91	4.16	1.36					

COBRATOXIN - CALF PEPTIDE COMPLEX

Solutions were prepared in which 0.5-1.0 equivalent of calf peptide was added to cobratoxin. Fig. 6 shows the spectrum in the aromatic region. Comparison of Figs. 2, 5, and 6 is instructive. It is noted that the aromatic resonances of the cobratoxin appear to be very little affected by the complex formation with the peptide. However the signals from the peptide are strongly affected. Most notably, the Y190δ and Y190ε resonances are shifted downfield to 7.20 and 6.89 respectively. Comparison of the COSY spectra of the mixture with that of neat cobratoxin allows a number of shifted resonances from the peptide to be identified. However, some COSY cross-peaks, especially those belonging to W184 and W187 are missing. It is presumed that intermediate exchange rates coupled with large chemical shifts are responsible,

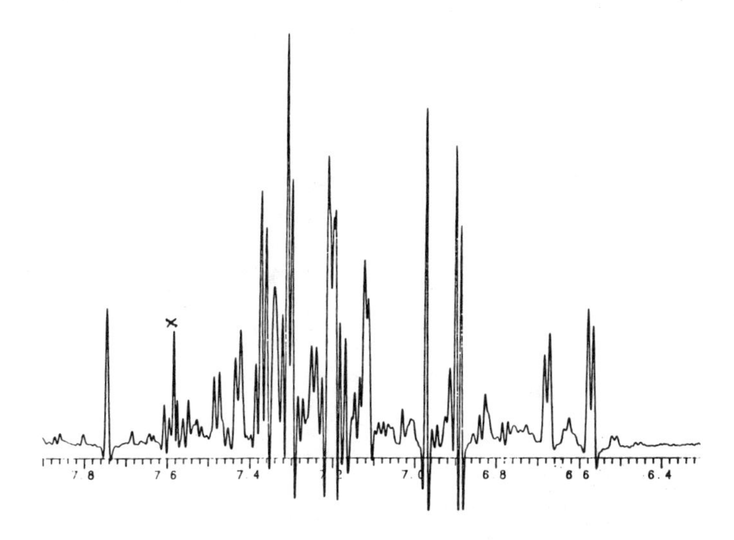

FIGURE 6. Aromatic region, 620 MHz spectrum of cobratoxin + 0.5 eq. calf peptide in D_2O, pH 3.4, $23^{o}C$.

but it is not excluded that the binding may be heterogeneous, and that the resonances are small and multiple.

That the basic structure of the cobratoxin is maintained intact may be ascertained by comparing the down-field shifted α-protons in the pleated sheet region (Fig. 7).

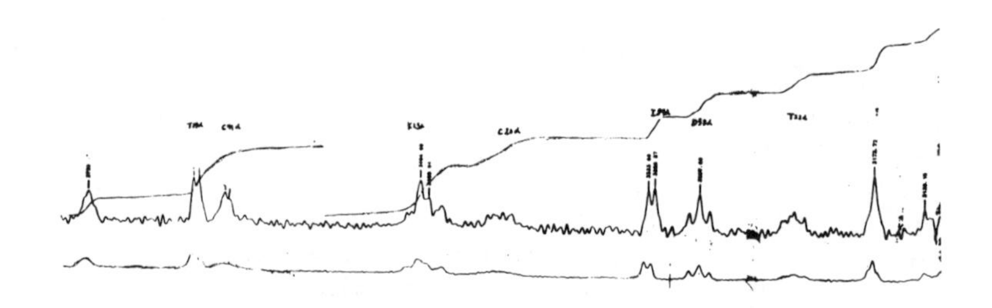

FIGURE 7. Comparison of spectra of β-sheet α- protons in cobratoxin(above) and cobratoxin + peptide (1:1) at pH 7.1, 23°C .

TABLE 2.

1-D AND 2-D DIFFERENCE SPECTROSCOPY YIELDS FOLLOWING CHANGES ON ADDITION OF PEPTIDE:

IN COBRATOXIN

V52γ	+0.01 PPM
V48γ	-0.01
T47γ	-0.06
F29 δ	+0.05
R36 δ	+0.10
K49,W25	~0
D27 ,D38 β	+0.01

IN PEPTIDE

E180 γ	+0.02
R182 δ	-0.04
W184(7)	-0.06
W187(7)	-0.10
V188 γ,γ'	+0.20,+0.30
Y191 δ	+0.12
Y191 ε	+0.14

Close examination of the full-range 1-D and 2-D spectra show the changes listed in Table 2. We interpret these changes as follows:

The shifts of T47γ, V48γ and V52γ, located in the "rail" indicate a displacement of the rail, with the side-chains headed out into the solution, as the tryptophan enters the cleft. The shift of F29δ reflects interaction with S181 of the peptide. E180 and R182 of the peptide form salt bridges with R33 and D27 of the cobratoxin, which flank F29. The shifts of D27β, D38β, E180γ and R182δ are consistent with the salt bridge formation, as originally postulated. The large shifts of V188γ and Y190δ and ε protons reflect the break-up of the peptide hydrophobic structure. The V188γprotons adopt a normal shift value, and are probably not involved directly in the binding.

Finally, a search was made for intermolecular NOES between peptide and cobratoxin. Only one has been able to be identified with fair confidence, to date. The effect is shown in Fig. 8 and is between W25(2) of the cobratoxin and the δ protons of Y190. Examination of models suggests that the tyrosine side chain can lie in the exit groove between residues C26 and G51, with a hydrogen bond to the CO of C26. This would place one of its δ protons close to W25(2).

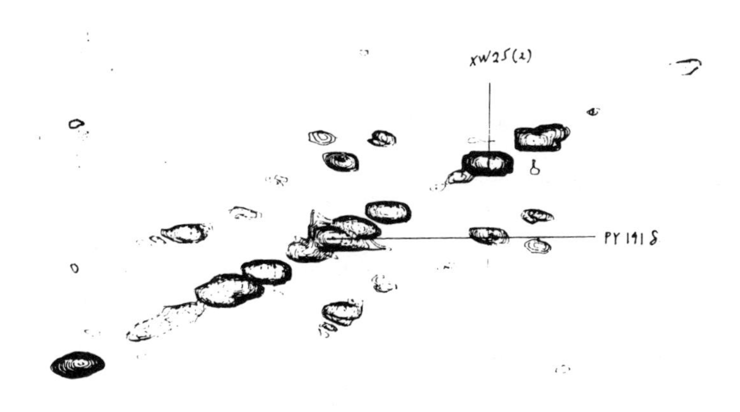

FIGURE 8. COSY (aromatic region) of cobratoxin + peptide (2:1),pH 6.3, 23^{0}C .

In summary the evidence supports the originally proposed binding model, and suggests that the bound peptide chain has its carbonyl terminus exiting past W25-C26.

ACKNOWLEDGEMENTS

We are very grateful for the great interest and work assisting with this problem provided by Professor Josef Dadok.

REFERENCES

1. Low BW, Corfield PWR, (1987) Acetylcholine Receptor: α-Toxin Binding Site-Theoretical and Model Studies. Asia Pacific Journal of Pharmacology 2:115.

2. Radding W, Corfield PWR, Levinson LS, Hashims GA, Low BW, (1988) α-Toxin Binding to Acetylcholine Receptor α179-191 Peptides Intrinsic-Fluorescence Studies. FEBS Letters 231:212.

3. Kondakov VI, Arsen'ev AS, Pluzknikov KA, Tsetlin VI, Bystrov VF, Ivanov VT (1984) 2-D NMR Investigation of the Conformational Features of Toxin 3 from the Venom of the Cobra _Naja naja siamensis_. Bioorg. Khim. 10:1606.

4. Walkinshaw MD, Saenger W. Maelicke A. (1980) Three-dimensional Structure of the Long Neurotoxin from Cobra Venom. Proc. Nat. Acad. Sci. USA 77:2400.

5. Karlsson E, (1979) Chemistry of Protein Toxins. In Lee CY "Snake Venoms". Handbook of Experimental Pharmacology, vol. 52. Berlin: Spring Verlag, p. 159.

6. Lee CY, (1972) Classification of Polypeptide Toxins from Elapid and Sea Snake Venoms According to their Pharmacological Properties and Chemical Structures. J. Formosan Med. Assoc. 71:311.

7. Changeux J-P, Kasai M, Lee Cy (1970) Use of a Snake Venom Toxin to Characterize the Cholinergic Receptor Protein. Proc. Nat. Acad. Sci. USA 67:1241.

8. Reynolds JA, Karlin A (1978) Molecular Weight in Detergent Solution of Acetylcholine Receptor from Torpedo California. Biochem. 17:2035.

9. Raftery MA, Hunkapiller MW, Strader CD, Hood LE (1980) "Acetylcholine Receptor: Complex of Homologous Subunits" Science 208:1454.

10. Kondakov VI, Arsen'ev AS, Ukin Yu N, Karlsson E, Gurevich AZ, Tsetlin VI, Bystrov VF, Ivanov VT (1984) NMR Study of the Spatial Structure of Toxin 3 of Naja naja siamensis Bioorg. Khim. 10:869.

11. Gross K-H, Kalbitzer HR (1988). Distribution of Chemical Shifts in ^{1}H Nuclear Magnetic Resonance Spectra of Proteins. J. Magn. Reson. 76:87.

Frontiers of NMR in Molecular Biology, pages 27-36

BUNGAROTOXIN BINDING TO ACETYLCHOLINE RECEPTOR-DERIVED SYNTHETIC PEPTIDES ANALYZED BY NMR[1]

Edward Hawrot, Kimberly L. Colson, Ian M. Armitage*, and G-Q. Song

Department of Pharmacology, and *Departments of Molecular Biophysics and Biochemistry and *Diagnostic Radiology, Yale University School of Medicine, New Haven, CT 06510 USA

ABSTRACT Amino acid residues 173 to 204 on the α-subunit of the nicotinic acetylcholine receptor form a major determinant of the ligand binding site, as has been shown by α-bungarotoxin binding to synthetic peptides corresponding to portions of this sequence. ^{1}H-nuclear magnetic resonance (NMR) spectroscopy has allowed us to probe the molecular details of the interactions that mediate the binding of α-bungarotoxin to these small synthetic peptides. NMR analysis of the aromatic region suggests that binding associated conformational changes occur both in α-bungarotoxin and in the receptor prototopes. Histidine residues on both are shown to be sensitive to binding-induced conformational changes, an observation that may indicate that these residues are directly involved in binding. Additionally, the NMR studies indicate that the microenvironment of the evolutionarily conserved tryptophan (TRP28) in α-bungarotoxin is affected upon interaction with the synthetic peptides strongly suggesting that the tryptophan side chain is directly involved in receptor recognition.

---

[1]This study was supported by NIH GM32629 (EH), DK18778 (IMA), the Muscular Dystrophy Association (EH), and instrumentation grants from NIH and NSF. EH is an Established Investigator of the American Heart Association.

INTRODUCTION

The determination of the primary amino acid sequences encoding the four subunits of the nicotinic acetylcholine receptor (AChR) (1-4), has opened the way towards the elucidation of functionally important structures within the AChR such as the protein domain comprising the acetycholine binding site. Attempts to localize the ligand binding site (10-17) have been facilitated by the use of the snake curaremimetic neurotoxins, which compete with acetylcholine with high affinity. Many of these neurotoxins are structurally well characterized (5-9) and in particular, both the X-ray crystallographic and the NMR-derived solution structure of α-bungarotoxin (BGTX) have been determined (18-21). We show here that the binding of BGTX with submicromolar affinity to a small synthetic peptide derived from the AChR sequence (22-25) offers an attractive, simplified system for investigating an important receptor-ligand interaction. Such small peptides, that correspond in sequence to an active region of a protein and that mimic the functional activity of that region, have recently been termed "prototopes" (26).

METHODS

All ^{1}H-NMR spectra were acquired at 500 MHz on a Bruker AM-500 NMR Spectrometer. Samples were prepared in D_2O at pH4 to increase solubility. Circular dichroism (CD) and intrinsic fluorescence studies of binding induced conformational changes indicate that at pH4, BGTX continues to bind to the appropriate synthetic peptides (22,23, and manuscript in preparation). For the formation of the complex between the 18mer and BGTX, equal amounts (246 μM) of the two were added together and the mixture was incubated overnight prior to NMR analysis. The pH of the solution was checked after complex formation and found to be unchanged. Presaturation of the residual HOD peak was used in the acquisition of all one-dimensional spectra. For the variable temperature NMR experiments, the temperature of the samples was varied in 5°K increments and allowed to equilibrate for at least 45 minutes prior to data acquisition. Referencing was external and to the C4 proton of the BGTX HIS4 at 6.51 ppm in a 246 μM BGTX solution at 298°K and pH4 (19-21). In this study we have used short synthetic peptides because of their increased solubility compared to the 32-mer (α-residues 173-204) previously used to demonstrate toxin binding to synthetic peptides. These more soluble, shorter peptides retain a considerable degree of toxin binding activity (K_D in the μM range) (12,22). Synthetic peptides were prepared and purified by reverse-phase HPLC in the Protein Chemistry Facility at Yale University School of Medicine.

RESULTS AND DISCUSSION

Our initial approach towards determining which specific residues are involved in binding has been to examine the one-dimensional proton NMR spectrum of both BGTX and the synthetic 18mer peptide (α residues 181-198) and to compare these individual spectra with the spectrum obtained from the 1:1 complex. Our attention has focussed on the aromatic region of the NMR spectrum because of the relative simplicity of the spectrum in this region. Although only the histidine C2 resonance of the 18mer will be discussed in this report, the other proton resonances of the free 18mer have been assigned by standard techniques (manuscript in preparation). As is shown in Fig. 1a, the single resonance for the C2 proton of HIS186 in the 18mer (acetyl-$_{181}$YRGWKHWVYYTCCPDTPY$_{198}$-amide) was identified at 8.49 ppm. The corresponding region of the spectrum for BGTX, determined under identical conditions, is shown in Fig. 1b. The labeled resonances were identified by comparison with previous reports (6,19-21). Upon formation of the 1:1 complex by addition of equimolar amounts of α-bungarotoxin to the 18mer (Fig. 1c), at least three groups of resonances appear to be noticeably affected: 1) the C2 resonances of HIS186 in the 18mer and HIS4 in BGTX, 2) the C4 resonance of HIS4 in BGTX and 3) the aromatic resonances of TRP28 in BGTX and of the two tryptophan residues in the 18mer.

A comparison of Figs. 1a, 1b, and 1c, shows that several sharp resonances of varying intensities appear in the spectrum of the complex with chemical shifts from 8.67 ppm to 8.43 ppm (five peaks in Fig. 1c marked with

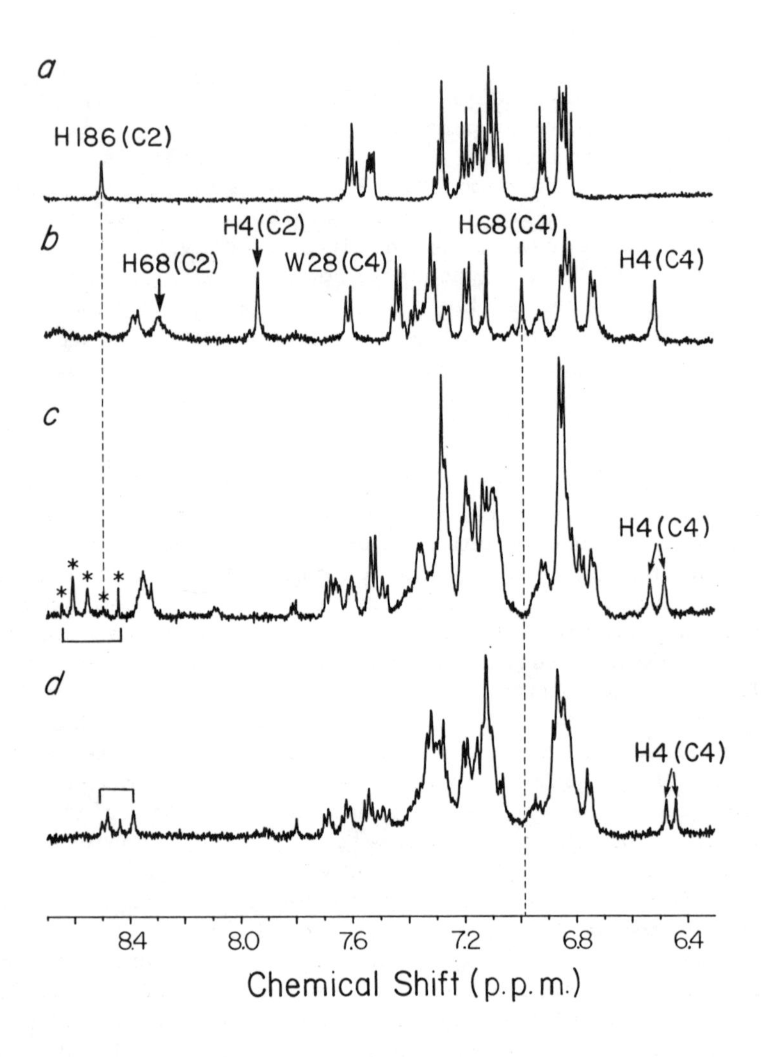

Figure 1. The downfield region of the ^{1}H NMR spectrum of *a*, the 18mer (246 μM), and *b*, BGTX (246 μM). *c*, The ^{1}H NMR spectrum for a 1:1 mixture of 18mer and BGTX at 298°K, and *d*, at 338°K. The populations of the two peaks corresponding to the C4 proton of HIS4 are 52%:48%.

asterisks). One of the peaks, as indicated in Fig. 1c, corresponds to the C2 peak of HIS186 prior to addition of BGTX and thus may represent a small amount of residual peptide that is not bound to BGTX. The other peaks that appear in this region of the complex (Fig 1c) suggest the co-existence of several different conformations surrounding the C2 proton of HIS186. If this explanation were true, then the alternative conformers, in slow exchange with one another, would be expected to exchange more rapidly as the temperature is raised to surmount the energy barrier between the different states. We performed variable temperature NMR experiments to test this hypothesis. A partial collapse in the distance separating the HIS186 C2 signals was observed upon raising the temperature to 338°K (Fig. 1d). At the elevated temperature, the peaks moved closer together, resonating over a chemical shift range of 0.11 ppm at 338°K as compared to the 0.23 ppm seen at 298°K (Fig. 1c). Thus the observed temperature-dependent partial coalescence of the multiple peaks associated with the C2 resonance of HIS186 is consistent with slow chemical exchange among a number of distinct conformers.

In addition to the marked changes seen with the easily identifiable HIS residue in the 18mer, the aromatic protons of both of the HIS residues in BGTX also exhibited significant spectral changes upon complex formation. The resonance due to the C2 proton of HIS4 of BGTX was shifted and severely broadened upon complex formation (Fig. 1c). In an attempt to determine whether the change in the C2 resonance of HIS4 was due to a marked shift in the resonance upon complex formation, a mixture containing a 1:2 ratio of peptide:BGTX was prepared. The resonance due to the unbound BGTX was clearly present in this mixture but at a reduced intensity as expected. We could not detect any new resonances in the same region as being due to a shift of the C2 resonance in the bound form of BGTX. We conclude, therefore, that it is likely that the resonance of the C2 proton of HIS4 is broadened into the baseline upon binding rather than being shifted completely out of this region of the spectrum. We could rule out broadening due to immobilization via gross mass aggregation since many other BGTX resonances in other regions of the spectrum remained unaltered (Fig. 1c). Variable temperature NMR experiments, which again would be expected to produce coalescence of exchangeable resonances, also failed to localize or sharpen further the C2 resonance of HIS4 in the complex, consistent with the observed effect being due either to structural immobilization or chemical exchange among multiple conformers. The behavior of the C4 resonance of HIS4 in the complex argues strongly for chemical exchange among conformers as described below.

The C4 proton of HIS4 resonates as a singlet at 6.51 ppm in the unbound form (Fig. 1b), but in the complex formed with the 18mer, two resonances of almost equal intensity and linewidth appear at 6.54 and 6.48 ppm (Fig. 1c). At high temperature (338°K), the separation between the two resonances was reduced to $\Delta\delta$=0.03 ppm (Fig. 1d) from $\Delta\delta$=0.05 ppm at 298°K. This reduction in the separation of the peaks with increased temperature argues that these two resonances correspond to two interconvertible conformers of HIS4 in the complex. The observed alterations in both the C4 and C2 protons of HIS4 indicate that these imidazole ring protons are sensitive indicators of the formation of the bound complex. Furthermore, the apparent slow exchange between two major conformers as evidenced by the formation of two resonances corresponding to the C4 proton of HIS4 is completely consistent with an exchange-mediated broadening of the C2 resonance on the same residue in the complex. Since the C2 resonance of histidines is usually more sensitive to environmental factors and displays a much broader range of chemical shifts (e.g., ref. 6), the exchange between two conformers in the complex could readily result in a broadening of the C2 resonance while producing a double set of resonances for the C4 proton of HIS4.

In addition to HIS4, the other histidine in BGTX (HIS68) is also sensitive to complex formation since the C4 and C2 resonances of this residue are each displaced in the complex (Fig. 1b,c). Together, these observations indicate that all three of the histidines in the complex of 18mer with BGTX are affected upon binding. The observed displacements may reflect the direct involvement of some or all of the histidines in binding or they may result from indirect conformational changes coincident with binding. Additional experiments would be required to determine whether HIS residues are directly involved in intermolecular recognition in this system. Bothner-By and coworkers have been examining the interaction of an overlapping peptide sequence with the related neurotoxin, α-cobratoxin (27). In α-cobratoxin, the two HIS residues (4 and 68) that we have focussed on in BGTX are both replaced by phenylalanines perhaps suggesting that the aromatic properties of the HIS residues may be an important feature in this region of the neurotoxin structure.

The resonance of the C4 proton in TRP28 was also affected in the 18mer-complex (Fig. 1b,c). The analysis of the 18mer-complex with simple one-dimensional ^{1}H NMR is somewhat complicated, however, because of the presence of two TRP residues in the 18mer. Similar studies with a 12mer (α-185-196) peptide (13,22) containing only one tryptophan residue indicate that the TRP28 C4 resonance is broadened into the baseline (23,24, and manuscript in preparation). These observations demonstrate that TRP28 in BGTX is clearly affected upon formation of the bound complex. In addition, the C4 and C7 proton resonances of the trypophan residues in the synthetic peptides appear to be shifted and slightly broadened in the complex. In other studies, we found that binding of the 12mer to BGTX resulted in a large increase in tryptophan-associated fluorescence and large changes in circular dichroism in the near UV region (24,25, and manuscript in preparation). Taken together, these observations are consistent with the involvement of tryptophan residues in binding. Using different peptides from the corresponding region in the calf AChR, Low and coworkers have observed fluorescence quenching upon binding of the peptides to α-cobratoxin (28).

The NMR studies reported here suggest an important functional role for TRP28 in the recognition by BGTX of the ligand binding domain on the AChR. The importance of TRP28 in neurotoxicity has been suggested previously based on the observation that TRP28 is 1) highly conserved within the entire family of curaremimetic neurotoxins (5,7,8), and 2) is situated in a central location within the crystal structure (5,7-9,18). Structural studies on the neurotoxins suggest that a hydrophobic cleft surrounds TRP28, and one model proposes that this cleft may interact with a tryptophan on the AChR (9,29).

Our results provide a direct demonstration that TRP28 in BGTX is markedly affected upon formation of the bound complex with synthetic peptide prototopes. These findings suggest that TRP28 may play a similar role in recognition and binding to the AChR. Additional studies involving two-dimensional NMR analysis of the bound complexes formed with the synthetic peptides will be required to determine the detailed sites of protein-protein contact. Together with distance geometry computational methods and molecular modeling, such studies should identify the receptor residues interacting with TRP28. Similarly, additional two-dimensional ^{1}H NMR investigations should reveal whether the histidine residues are directly involved in binding or whether the induced chemical shifts are due to conformational rearrangements coincident with complex formation.

If conformational changes are responsible for the HIS resonance displacements in bound BGTX, then it would appear that, despite its proposed structural rigidity (6), significant conformational changes must occur in BGTX upon interaction with receptor. Although much is known of the structure of the curaremimetic neurotoxins (5-9), including the X-ray crystal structure (18) and the NMR solution structure of BGTX (19-21), little is known of the final conformation achieved upon binding to the AChR. Further NMR studies on the structure of the complex formed with the synthetic peptides should provide valuable information on this question.

The present study has utilized synthetic peptides (prototopes) comprising a binding site domain on the AChR to identify specific amino acid residues involved in binding. NMR spectroscopy offers the prospect of elucidating the complete three-dimensional solution conformation of the bound complex and would thus identify those residues responsible for ligand binding. This general strategy towards elucidating the structural requirements of receptor binding sites should have a direct bearing on the future rational design of drugs and similar approaches may be useful in other recognition systems to test putative binding domains that may be predicted from amino acid sequence information.

Synthetic peptides are being increasingly used in biological studies to correlate structure with function and hence there is great interest in determining the native conformation of such peptides (30). In other studies, we have shown that the 12mer may contain a small amount of ordered structure based on differences in the rate of exchange of some peptide amide protons in D_2O at elevated temperatures (31). In order to evaluate the importance of such structural constraints on the mode of binding, the detailed structures of the peptides will have to be analyzed and compared. We are presently pursuing studies aimed at determining the intrinsic structure of the isolated synthetic peptides. We also are investigating the time course of binding in order to identify possible intermediates in the formation of the complex. Such studies may provide additional insights into the mode of binding and the molecular basis for the high-affinity interaction that we observe in this important model system.

ACKNOWLEDGEMENTS

We thank Dr. Thomas Lentz for valuable discussions and Dr. Basus for providing us with revised sequence information and with the ^{1}H NMR chemical shift data on α-bungarotoxin prior to publication.

REFERENCES

1. Claudio T, Ballivet M, Patrick J, and Heinemann S (1983). Proc. Natl. Acad. Sci. 80:1111.
2. Noda M, Takahashi H, Tanabe T, Toyosato M, Kikyotani S, Furutani Y, Hirose T, Takashima H, Inayama S, Miyata T, and Numa S (1983). Nature 302:528.
3. Sumikawa K, Houghton M, Smith JC, Bell L, Richards BM, and Barnard EA (1982). Nucleic Acids Res. 10:5809.
4. Devillers-Thiery A, Giraudat J, Bentaboulet M, and Changeux J-P (1983). Proc. Natl. Acad. Sci. 80:2067.
5. Karlsson E (1979). "Snake venoms, Handbook of Experimental Pharmacology" Berlin: Springer-Verlag, 52:159.
6. Endo T, Inagaki F, Hayashi K, and Miyazawa T (1981). Eur. J. Biochem. 120:117.
7. Dufton MJ and Hider RC (1983). CRC Crit. Rev. Biochem. 14:113.
8. Allen M and Tu AT (1985). Molecular Pharmacology 27:79.

9. Low BW and Corfield PWR (1986). Eur. J. Biochem. 161:579.
10. Karlin, A (1980). Cell Surface Rev. 6:191.
11. Wilson PT, Lentz TL, and Hawrot E (1985). Proc. Natl. Acad. Sci. 82:8790.
12. Neumann D, Barchan D, Safran A, Gershoni JM, and Fuchs S (1986). Proc. Natl. Acad. Sci. 83:3008.
13. Neumann D, Barchan D, Fridkin M, and Fuchs S (1986). Proc. Natl. Acad. Sci. 83:9250.
14. Barkas T, Mauron A, Roth B, Alliod C, Tzartos SJ, and Ballivet M (1987). Science 235:77.
15. Gershoni JM (1987). Proc. Natl. Acad. Sci. 84:4318.
16. Ralston S, Sarin V, Thanh HL, Rivier J, Fox JL, and Lindstrom J (1987). Biochemistry 26:3261.
17. Mulac-Jericevic B and Atassi MZ (1986). FEBS Letters 199:68.
18. Love RA and Stroud RM (1986). Protein Engineering 1:37.
19. Basus VJ, Billeter M, Love RA, Stroud RM, and Kuntz ID (1988). Biochemistry 27:2763.
20. Basus VJ and Scheek RM (1988). Biochemistry 27:2772.
21. Kosen PA, Finer-Moore J, McCarthy MP, and Basus VJ (1988). Biochemistry 27:2775.
22. Wilson PT, Hawrot E, and Lentz TL (1988). Molecular Pharmacology 34:643.
23. Song G-Q, Armitage I, and Hawrot E. (1989). Biophys. J. 55:149a.
24. Shi Q-L, Colson KL, Lentz TL, Armitage IM, and Hawrot E (1988). Biophys. J. 53:94a.
25. Pearce SF and Hawrot E (1989). Biophys. J. 55:517a.
26. House C and Kemp BE (1987). Science 238:1726.
27. Bothner-By AA (1989). This volume.
28. Radding W, Corfield PWR, Levinson LS, Hashim GA, and Low BW (1988). FEBS Letters 231:212.
29. Low BW and Corfield PWR (1987). Asia Pacific J. of Pharmacology 2:115.
30. Dyson HJ, Cross KJ, Houghton RA, Wilson IA, Wright PE, and Lerner RA (1985). Nature 318:480.
31. Hawrot E, Lentz TL, Colson KL, and Wilson PT (1988). Synthetic peptides in the study of the nicotinic acetylcholine receptor. In "Current Topics in Membranes and Transport," New York: Academic Press, 33:165.

Frontiers of NMR in Molecular Biology, pages 37-49

NMR STUDIES OF THE FIRST ZINC FINGER DOMAIN FROM THE GAG PROTEIN OF HIV-1[1]

Michael F. Summers, Terri L. South, Bo Kim, and Dennis Hare

Chemistry Department, University of Maryland Baltimore County, Baltimore, MD 21228, and Hare Research Inc. (D.R.H.), 14810 216th Ave N.E., Woodinville, WA 98072

Modern NMR spectroscopic methods have been utilized to study zinc and cadmium adducts with an 18-residue synthetic peptide containing the amino acid sequence of the first zinc finger domain from the gag protein of HIV-1 (the causative agent of AIDS). The metal binding mode of the ^{113}Cd adduct was determined unambiguously using ^{1}H-^{113}Cd correlation NMR spectroscopy. The three-dimensional structure of the Zn^{2+} adduct was determined using a novel NMR/distance geometry/NOE-back calculation approach. The complex contains a potential nucleic acid interaction site consisting of a hydrophobic patch with adjacent amino-containing side chains.

INTRODUCTION

During assembly and budding stages of the retroviral life cycle, a gag polyprotein identifies and forms a complex with viral RNA (1,2). Without exception, retroviral gag proteins (and their proteolytically cleaved nucleic acid binding protein (NABP) products) contain either one or two conserved "zinc finger" regions (3) of the type, -C-X_2-C-X_4-H-X_4-C (C = cysteine, H = histidine, X = variable amino acid) (4-6). Although related sequences

[1]This work was supported by ACS Institutional Research Grant No. IN-147F, MD Cancer Program/University of Maryland (M.F.S.), the NOVA Pharmaceutical Corporation (M.F.S.), and by NIH grant GM35620-02 (D.R.H.).

found in DNA-binding proteins bind zinc tightly (7-12), experiments aimed at measuring the zinc content and the affinity of zinc for retroviral NABPs indicate that zinc binds weakly to these proteins (13,14), and this has led to the conclusion that zinc is not a structural component of at least one retroviral NABP (13). On the other hand, recent site directed mutagenesis experiments involving murine leukemia virus provide evidence that zinc binding is necessary for correct protein function, and indicate that retroviral zinc fingers function directly in retroviral RNA recognition processes (15).

We have prepared Zn^{2+} and Cd^{2+} adducts with the 18-residue peptide comprising the amino acid sequence of the first finger (residues 13 through 30) of NABP p7 from HIV-1 (16). The stable 1:1 metal adducts formed allowed for detailed investigations into the metal binding mode and backbone folding patterns (17).

METHODS

Data acquisition and processing.

NMR spectral measurements were made with a GE GN-500 (500 MHz, ^{1}H) spectrometer. Raw data were transferred via ethernet to clustered VAX8600 mainframe computers and processed using the FTNMR software package. References to pulse sequences and specific parameters follow. ^{1}H-^{113}Cd Heteronuclear spin-echo difference (HSED) spectroscopy (18): 15 mM $^{113}Cd(p7^{13-30})$; T = -5.0 °C; 5,000 Hz spectral width; 4K data size; 128 scans; 1.5 Hz Gaussian filtering. Heteronuclear multiple quantum coherence (HMQC) spectroscopy (19): 26 μs 90° ^{13}C pulse width; MLEV-16 modulated ^{13}C broad-band decoupling during the acquisition period; 2 x 256 x 1024 data matrix size; 48 scans (preceded by 2 dummy scans) per t_1 value; pulse delay = 1.0 s; 6 Hz Gaussian filtering and 90°-shifted squared sine bell filtering in the t_2 and t_1 dimensions, respectively. Homonuclear Hartmann-Hahn (HOHAHA) spectroscopy (20,21): 2 x 350 x 2048 data matrix size; 16 scans per t_1 value; pulse delay = 2.0 s, MLEV-17 mixing time = 33 ms, preceded and followed by 2.5 ms continuous wave trim pulses; 6 W rf power corresponding to 35 μs 90° pulse widths; 3 and 6 Hz Gaussian filtering in the t_2 and t_1 dimensions, respectively. Nuclear Overhauser effect (NOESY) spectroscopy (22): 2 x 512 x 2048 data matrix sizes; 16 and

64 scans per t_1 value for data obtained in D_2O and H_2O solutions, respectively; τ_{mix} values of 5, 50, 100, 300, and 500 ms; total recycle delay = 2.5 s (= 2.5 times the longest proton T_1 value); 6 Hz exponential filtering and 90°-shifted squared sine bell filtering in the t_2 and t_1 dimensions, respectively. Rotating frame Overhauser effect (ROESY) spectroscopy (20,23): τ_m = 85 ms; 40 µs 90° pulse width; all other parameters the same as for HOHAHA spectroscopy. Double quantum filtered (2QF) COSY (24): 2 x 512 x 2048 data matrix size; 32 scans per t_1 value; data were processed using squared sine bell filters in the t_1 and t_2 dimensions.

Distance geometry calculations.

Volume integrals of all resolved (and partially overlapping) cross-and auto-peaks were calculated using the FTNMR software package (25). Signal intensities observed in the NOESY data obtained with a mixing time of 100 ms were used to make initial interproton distance estimates. From these estimates, a list of upper and lower interproton distance constraints was prepared, which was used as input for the distance geometry program, DSPACE (25). Families of structures were generated using Zn-S(Cys) and Zn-N(His-N^3) distances of 2.3 and 2.1 Å, respectively (26). No bond angle constraints involving Zn were employed. Initial structures exhibited considerable backbone variations and often contained non-tetrahedral zinc coordination geometries.

These structures were then used to back-calculate theoretical 2D NOESY spectra using the program, BKCALC (25). The theoretical spectra generated with BKCALC contained a number of NOE cross-peaks not observed in the experimental spectra, and minimum internuclear distance constraints were added to the constraints list for the appropriate proton pairs. The above cycle was then repeated until the experimental and theoretical 2D NOESY spectra were visually identical. Structure refinements were made by adjusting the interproton distance constraints to obtain a match between the experimental and theoretical NOE build-up curves and auto peak decay curves.

RESULTS

Metal binding studies.

^{1}H NMR experiments indicate that the synthetic peptide ($p7^{13-30}$) forms 1:1 metal adducts that are highly stable in aqueous solution (pH 7, ambient T). Except for a few broad ^{1}H NMR signals observed in the spectrum of the ^{113}Cd adduct, (particularly the His-H^2 and -H^4 signals), the ^{1}H spectra of Zn($p7^{13-30}$) and ^{113}Cd($p7^{13-30}$) are similar in appearance. The broader signals of the ^{113}Cd adduct narrow considerably on cooling to -5 °C. Of the 17 backbone amide protons in Zn($p7^{13-30}$), 12 protons exhibit resolved multiplets (due to NH-CHα scalar coupling) in the ^{1}H spectrum obtained at 30 °C. The ^{113}Cd NMR spectrum of ^{113}Cd($p7^{13-30}$) exhibits a single resonance at δ 653 ppm, which is within the range expected for Cd bound predominantly by S^- donor ligands (27). The identities of the coordinated ligands were determined more precisely using ^{1}H-^{113}Cd heteronuclear spin-echo difference (HSED) NMR spectroscopy. In ^{1}H-^{113}Cd HSED spectra, only the ^{1}H signals of protons that are scalar coupled to ^{113}Cd are observed. The ^{1}H-^{113}Cd HSED spectrum of ^{113}Cd($p7^{13-30}$) (Fig. 1B) exhibits signals for both the His-H^2 and -H^4 protons, providing unambiguous evidence that the His imidazole is coordinated to ^{113}Cd via N^3. HSED correlation

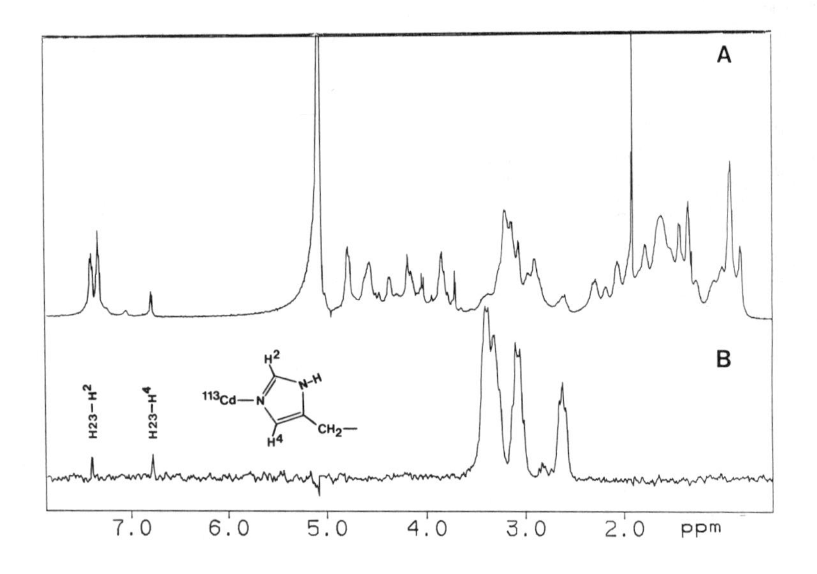

FIGURE 1. 500 MHz ^{1}H (A) and ^{1}H-^{113}Cd HSED (B) spectra of ^{113}Cd($p7^{13-30}$); T = -5.0 °C.

signals also observed for Cys(β) protons (2.5-3.5 ppm, presently unassigned) confirm the presence of Cd-Cys bonds.

^{1}H NMR signal assignments were made using the general approach outlined by Wuthrich (28). Data were obtained with sample temperatures of 30 °C in order to achieve maximum resolution of the backbone amide proton signals. A portion of the HOHAHA spectrum obtained for Zn($p7^{13-30}$) using a sample containing 90%H_2O/10%D_2O as solvent is shown in Fig. 2. Correlation signals are observed for all backbone amide protons except V(13) (terminal NH_3^+), A(25) and R(26). From these data, signals were grouped into individual "J-networks" (networks of scalar coupled spins), representing each of the amino acid spin systems. Many of the spin systems were assigned to generic amino acid types; for example, the amide signals at 7.93 and 8.43 ppm each gave HOHAHA correlation signals to two α-protons, indicating that these signals correspond to glycine residues. Generic assignments for residues with longer side chains were facilitated by comparison with 2QF-COSY data (not shown).

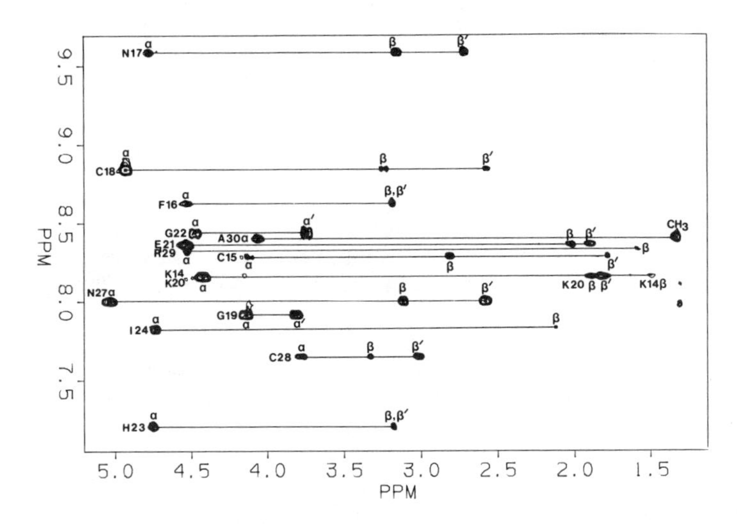

FIGURE 2. Region of the 2D HOHAHA spectrum of Zn($p7^{13-30}$) exhibiting connectivities between backbone NH, Hα, and Hβ protons.

In some cases, severe signal overlap in the 2QF-COSY spectrum for protons of neighboring methylene groups precluded assignment of β, γ, δ, and ε proton signals. These signals were differentiated using ^{1}H-^{13}C HMQC data, where signals due to geminal protons are readily identified (Fig. 3). For example, by spreading signals out in the ^{13}C dimension, differentiation of K(20) β, γ, and δ signals was possible.

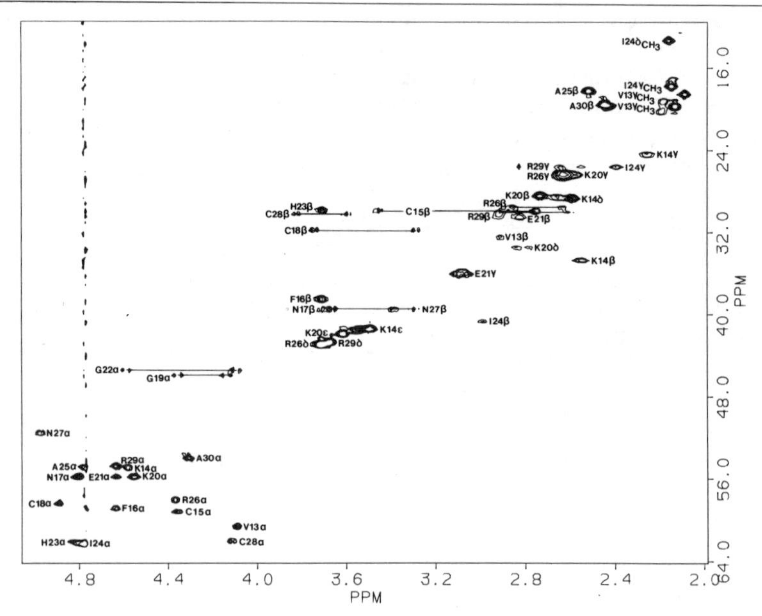

FIGURE 3. ^{1}H-^{13}C HMQC spectrum (upfield ^{1}H and ^{13}C regions) with assignments for Zn($p7^{13-30}$). Signals for some geminal protons are indicated by horizontal lines.

Once the protons had been grouped into individual J-networks, assignments to specific amino acids were made using 2D ROESY and NOESY NMR spectroscopy. The downfield region of the ROESY spectrum of Zn($p7^{13-30}$) is shown in Fig. 4. The longest stretch of sequential NH-NH connectivities was observed for a spin system that contained three sequential residues with CH_2-X side chains, which was followed by Gly and Lys residues. Based on the known sequence, these signals were assigned to F(16)-N(17)-C(18)-G(19)-K(20). An additional NH-NH connectivity between residues with CH_2-X side chains was

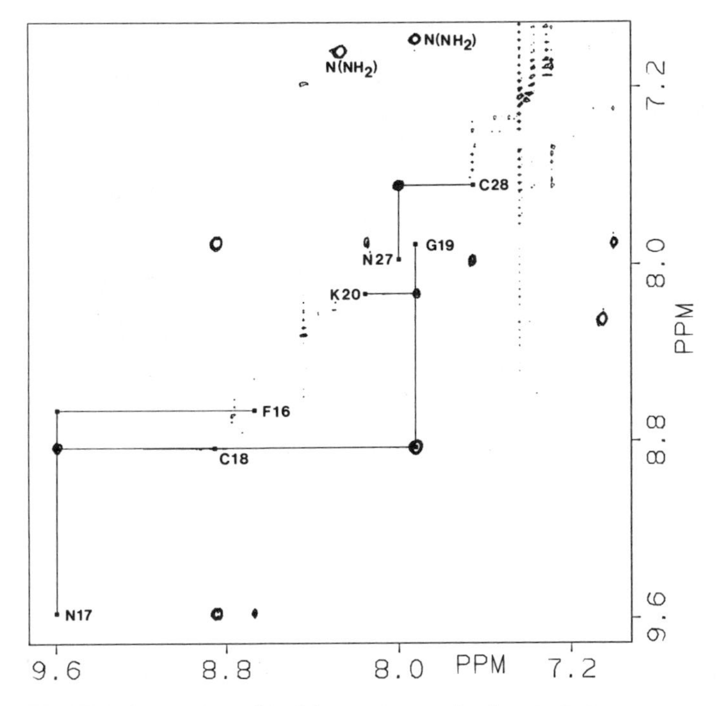

FIGURE 4. Downfield region of the ROESY spectrum of Zn(p7$^{13-30}$). Sequential NH-NH connectivities and amino acid assignments are indicated.

assigned as the N(27)-C(28) connectivity. Correlation signals for asparagine amide protons are also observed in Fig. 4.

Many additional interresidue connectivities were observed in the NH-CHα region of the Overhauser effect spectra. This region of the 2D NOESY data (τ_m = 300 ms) is shown in Fig. 5. Examples of sequential connectivities include K(14)-Hα to C(15)-HN, C(15)-Hα to F(16)-HN, F(16)-Hα to N(17)-HN, C(18)-Hα to G(19)-HN, G(19)-Hα to K(20)-HN, K(20)-Hα to E(21)-HN. ^{1}H NMR chemical shift assignments are summarized in Table 1.

Structural analysis of Zn(p7$^{13-30}$).

A family of structures was obtained for Zn(p7$^{13-30}$) using the final constraints list and randomly generated initial atomic coordinates. Each of the final structures

TABLE 1
^{1}H NMR CHEMICAL SHIFTS FOR Zn(p7$^{13-30}$)[a]

Residue	NH	CαH	CβH	Others
V(13)		3.73	2.01	γCH_3 0.82, 0.88
K(14)	8.17	4.45	1.47	γCH_2 1.06 δCH_2 1.55 εCH_2 2.87
C(15)	8.01	4.14	2.82 1.79	
F(16)	8.67	4.54	3.19	H^{2-6} 7.32-7.38
N(17)	9.58	4.78	2.72 3.17	Amide 7.07, 8.21
C(18)	8.83	4.92	2.57 3.24	
G(19)	7.93	3.81 4.13		
K(20)	8.17	4.42	1.84	γCH_2 1.51, 1.65 δCH_2 1.91 εCH_2 3.05
E(21)	8.38	4.55	1.91 2.04	γCH_2 2.23, 2.31
G(22)	8.43	3.76 4.47		
H(23)	8.30	4.15	1.79 2.83	H^2 7.43 H^4 6.87
I(24)	7.84	4.77	2.13	γCH_2 1.26 γCH_3 0.90 δCH_3 0.93
A(25)	8.43	4.14	1.42	
R(26)	8.22	4.15	1.61	γCH_2 1.93 δCH_2 3.19
N(27)	8.01	4.38	2.59 3.12	Amide 7.00, 7.91
C(28)	7.65	3.79	3.02 3.33	
R(29)	8.38	4.54	1.57 2.02	γCH_2 1.53 δCH_2 3.13
A(30)	8.72	4.07	1.34	

[a]Chemical shifts in ppm relative to internal H_2O (4.725 ppm). Sample conditions: 20 mM in Zn(p7$^{13-30}$); T = 30 ^oC; pH = 7.0.

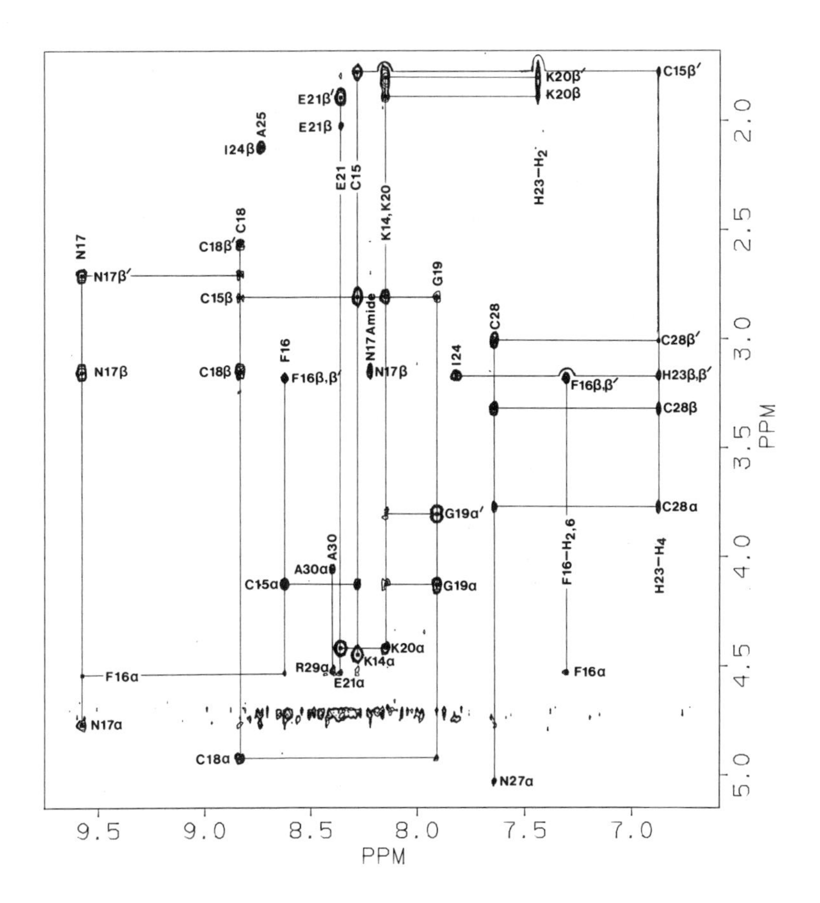

FIGURE 5. Portion of the 2D NOESY spectrum of $Zn(p7^{13-30})$.

refined to a total penalty (penalty = squared sum of the covalent and experimental bounds violations) of less than 0.45 $Å^2$, using 171 interproton distance constraints with bounds for the 63 observable distances of $\pm$ 0.1 Å or less. Measured values of $^3J(HC\alpha\text{-}NH)$ couplings were consistent with backbone torsion angles in the final structures (28). In no cases were NOE peaks observed that would be indicative of a dynamic process (29). Although not implicitly invoked, the final structures exhibited tetrahedral or near-tetrahedral Zn coordination geometries.

DISCUSSION

$Zn(p7^{13-30})$ exhibits a novel folding pattern that includes two Type-I tight turns within the 14-residue zinc binding domain. This folding is significantly different from that predicted (30) and found (12) for zinc finger complexes from DNA-binding proteins. A ribbon drawing showing the general folding pattern is shown in Fig. 6. With this folding, the phenyl group of the phenylalanine residue and the methyl groups of the isoleucine residue are adjacent to each other and make up a hydrophobic patch on the peptide surface. These residues are at conservatively substituted sites (based on comparisons of amino acid sequences of the zinc finger domains of known retroviral *gag* proteins) that have been implicated recently in genomic RNA recognition (15). The 2D NOE back-calculation/distance geometry approach utilized in this work provides a powerful method for studying the structural features of peptides and small proteins. We expect this approach will find wide use in other applications of NMR in molecular biology.

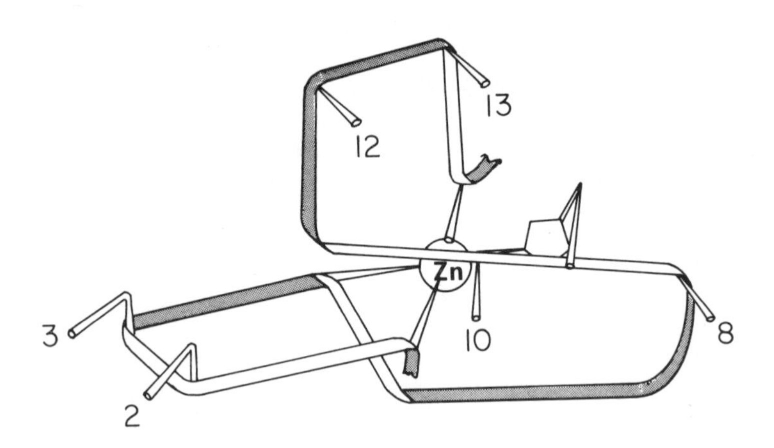

FIGURE 6. Drawing showing the general backbone folding observed in $Zn(p7^{13-30})$. An internal numbering scheme is used with the first Cys residue defined as C(1). Sites that are conservatively substituted in retroviral *gag* proteins are indicated by numbers.

ACKNOWLEDGMENTS

Technical assistance from Dr. H. R. Summers and J. Suess (UMBC), and helpful discussions with Drs. R. L. Karpel (UMBC) and L. E. Henderson (Frederick Cancer Research Facility) are gratefully acknowledged.

REFERENCES

1. Weiss R, Teich N, Varmus H, Coffin J, (eds): "RNA Tumor Viruses", 2nd Ed., Vols. 1 and 2, New York: Cold Spring Harbor Laboratory, Cold Spring Harbor, p 1108.
2. Bolognesi DP, Montelaro RC, Frank H, Schafer W, (1978). Assembly of type C oncornaviruses: a model. Science 199:183.
3. Berg JM (1986). Potential metal binding domains in nucleic acid binding proteins. Science 232:485.
4. Henderson LE, Copeland TD, Sowder RC, Smythers GW, Oroszlan S (1981). Primary structure of the low molecular weight nucleic acid binding proteins of murine leukemia viruses. J Biochem 256:8400.
5. Copeland TD, Morgan MA, Oroszlan S (1984). Complete amino acid sequence of the basic nucleic acid binding protein of feline leukemia virus. Virology 133:137.
6. Karpel RL, Henderson LE, Oroszlan S (1987). Interactions of retroviral structural proteins with single-stranded nucleic acids. J Biol Chem 262:4961.
7. Miller J, McLachlan AD, Klug A (1985). Repetitive zinc-binding domains in the protein transcription factor IIIA from Xenopus oocytes. EMBO J 4:1609.
8. Diakun GP, Fairall L, Klug A (1986). EXAFS study of the zinc-binding sites in the protein transcription factor IIIA. Nature 324:698.
9. Giedroc DP, Keating KM, Williams KR, Konigsberg WH, Coleman JE (1986). Gene 32 protein, the single-stranded DNA binding protein from bacteriophage T4, is a zinc metalloprotein. Proc Natl Acad Sci USA 83:8452.
10. Wingender E, Seifart KH (1987). Transcription in eukaryotes- the role of transcription complexes and their components. Angew Chem Int Ed Engl 26:218.
11. Frankel AD, Berg JM, Pabo CO (1987). Metal-dependent

folding of a single zinc finger from transcription factor IIIA. Proc Natl Acad Sci USA 84:4841.

12. Parraga G, Horvath SJ, Eisen A, Taylor WE, Hood L, Young ET, Klevit RE (1988). Zinc-dependent structure of a single-finger domain of yeast ADR1. Science 241:1489.
13. Jentoft JE, Smith LM, Fu X, Johnson M, Leis J (1988). Conserved cysteine and histidine residues of the avian myeloblastosis virus nucleocapsid protein are essential for viral replication but are not "zinc-binding fingers". Proc Natl Acad Sci USA 85:7094.
14. Schiff LA, Nibert ML, Fields BN (1988). Charactarization of a zinc blotting technique: evidence that a retroviral gag protein binds zinc. Proc Natl Acad Sci USA 85:4195.
15. Gorelick RJ, Henderson LE, Hanser JP, Rein A (1988). Point mutants of Moloney murine leukemia virus that fail to package viral RNA: Evidence for specific RNA recognition by a "zinc finger-like" protein sequence. Proc Natl Acad Sci USA 85:8420.
16. South TL, Kim B, Summers MF (1989). ^{113}Cd NMR studies of a 1:1 Cd adduct with an 18-residue finger peptide from HIV-1 nucleic acid binding protein, p7. J Am Chem Soc 111:395.
17. Summers MF, Hare DR, South TL, Kim B, manuscript in preparation.
18. Freeman R, Mareci TH, Morris GA (1981). Weak satellite signals in high-resolution NMR spectra: separating the wheat from the chaff. J Magn Reson 42:341; Cohen JS, Chen C-W, Bax A (1984). Selective observation of phosphate ester protons by proton {phosphorus-31} spin-echo difference spectroscopy. J Magn Reson 59:181.
19. Bax A, Subramanian S (1986). Sensitivity-enhanced two-dimensional heteronuclear shift correlation NMR spectroscopy. J Magn Reson 67:565.
20. Bax A, Sklenar V, Clore GM, Gronenborn AM (1987). Water suppression in two-dimensional spin-locked nuclear magnetic resonance experiments using a novel phase-cycling procedure. J Am Chem Soc 109:6511.
21. Braunschweiler L, Ernst RR (1983). Coherence transfer by isotropic mixing: application to proton correlation spectroscopy. J Magn Reson 53:521; Davis DG, Bax A (1985). Assignment of complex ^{1}H NMR spectra via two-dimensional homonuclear

Hartmann-Hahn spectroscopy. J Am Chem Soc 107:2820.
22. A modified version (to be published) of the pulse sequence described by Bax and co-workers (Sklenar V, Bax A (1987). Spin-echo water suppression for the generation of pure-phase two-dimensional NMR spectra. J Magn Reson 74:469; Sklenar V, Brooks BR, Zon G, Bax A (1987). Absorption mode two-dimensional NOE spectroscopy of exchangable protons in oligonucleotides. FEBS Let 216:249) was employed.
23. Bothner-By A, Stephens RL, Lee JT, Warren CD, Jeanloz RW (1984). Structure determination of a tetrasaccharide: transient nuclear Overhauser effects in the rotating frame. J Am Chem Soc 106:811. Davis D, Bax A (1985). Separation of chemical exchange and cross relaxation effects in two-dimensional NMR spectroscopy. J Magn Reson 63:207.
24. Rance M, Sorensen OW, Bodenhausen G, Wagner G, Ernst RR, Wuthrich K (1983). Improved spectral resolution in COSY ^{1}H NMR spectra of proteins via double quantum filtering. Biochem Biophys Res Commun 117:479.
25. Hare Research, Inc.
26. Distance estimates were made from EXAFS data obtained for transcription factor IIIA. See ref 8.
27. Summers MF (1988). ^{113}Cd NMR spectroscopy of coordination compounds and proteins. Coord Chem Revs 86:43 and references therein.
28. Wuthrich K (1986). "NMR of Proteins and Nucleic Acids" New York: John Wiley and Sons.
29. No inconsistent internuclear distances involving three or more nuclei were observed.
30. Berg JM (1988). Proposed structure for the zinc-binding domains from transcription factor IIIA and related proteins. Proc Natl Acad Sci USA 85:99.

Frontiers of NMR in Molecular
Biology, pages 51-61

CONFORMATIONS OF ENZYME-BOUND LIGANDS BY ISOTOPE-EDITED 2D NMR TECHNIQUES

Stephen W. Fesik, Jay R. Luly, John W. Erickson,
Celerino Abad-Zapatero, and Erik R.P. Zuiderweg

Pharmaceutical Discovery Division, Abbott Laboratories
Abbott Park, Illinois 60064

ABSTRACT Methods are described for providing detailed structural information on large enzyme/inhibitor complexes using NMR spectroscopy. The methods involve the use of isotopically labeled ligands to simplify NMR spectra of large molecular complexes. In one approach, only those protons which are attached to isotopically labeled nuclei (^{13}C, ^{15}N) are selectively detected using isotope-editing techniques. From a series of isotope-edited proton NMR spectra, amide proton exchange rates were measured for a tightly bound ^{15}N-labeled tripeptide inhibitor of porcine pepsin (IC_{50} = 1.7 X 10^{-7}M). In addition, using these techniques the backbone and side-chain conformations (at the P_2 and P_3 sites) of the inhibitor when bound to the enzyme were determined, and structural information on the active site was obtained. In another approach, spectral simplification was achieved by taking the difference of 2D NOE spectra of an enzyme/protonated inhibitor complex with that of an enzyme/deuterated inhibitor complex. Using this method, NOES involving the protons of the inhibitor could easily be identified.

INTRODUCTION

Experimentally determined three-dimensional structures of enzyme/inhibitor complexes are playing an increasingly important role in the design of novel enzyme inhibitors with therapeutic utility (1,2). Thus far, this structural information has come from X-ray crystallography;

however, in principle, NMR spectroscopy could also be used to provide structural information on enzyme/inhibitor complexes helpful in designing more effective inhibitors. Indeed, from nuclear Overhauser effect (NOE) measurements which indicate the proximity of protons in space (3), three-dimensional structures of molecules up to a molecular weight of 10 kDa have been determined (4,5). However, enzyme/inhibitor complexes are typically larger and not amenable to study by conventional NMR techniques. In order to overcome some of the limitations associated with proton NMR studies of large molecules, experimental approaches have been proposed (6-14). One approach relies on the simplification of proton NMR spectra by the selective observation of protons attached to isotopically labeled nuclei (e.g. ^{13}C, ^{15}N). Simplification of proton NMR spectra through isotope-editing is achieved by taking the difference of proton NMR spectra acquired with or without a radiofrequency field (180° pulse)

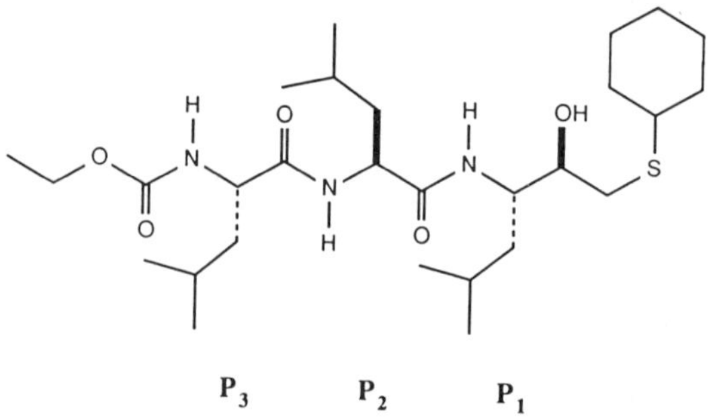

FIGURE 1. Structure of the pepsin inhibitor [Source: S.W. Fesik et. al. (16)].

applied at the frequency of the isotopically labeled nucleus (6,7). Using these techniques, we have measured the amide proton exchange rates for a tightly bound inhibitor (Figure 1) of porcine pepsin (15). In addition, isotope-editing allowed us to determine the conformation of the enzyme-bound pepsin inhibitor as well as to obtain structural information on the active site (16).

In another approach, NOEs involving the protons of the inhibitor were identified by subtracting two-dimensional NOE spectra of two enzyme/inhibitor complexes prepared with either a protonated or deuterated inhibitor.

METHODS

Porcine pepsin A was obtained from Sigma Chemical Co. and the labeled tripeptide pepsin inhibitor was synthesized using isotopically labeled (^{15}N, ^{13}C, or ^{2}H) leucine (Cambridge Isotopes) as previously described for the preparation of related inhibitors (17). The NMR samples were prepared by rapidly mixing a DMSO-d_6 solution (0.025 mL) of the isotopically labeled inhibitor into a $H_2O/^2H_2O$ (9/1) or a ^2H_2O solution of pepsin at pH = 3.7. The small amount (5%) of DMSO used to solubilize the inhibitor did not have any deleterious effect on the ability of the tripeptide to inhibit pepsin (IC_{50} = 1.7 X 10^{-7} M).

NMR experiments were performed on a General Electric GN500 or Bruker AM500 NMR spectrometer. The pulse sequence used in the isotope-edited 2D NOE experiments consisted of a conventional 2D NOE experiment in which a proton spin-echo pulse sequence (90°-τ-180°-τ) was substituted for the last proton pulse (14). In addition, a 180° pulse was applied on alternate scans at the frequency of the X-nucleus (^{13}C or ^{15}N) concurrent with the proton 180° refocusing pulse of the spin-echo sequence. By subtraction of the data collected with or without a 180° X-nucleus pulse, only those protons attached to the isotopically labeled nuclei and their dipolar coupled partners were observed. Decoupling of the X-nucleus was achieved during the t_1 and acquisition (t_2) periods with a MLEV-64 pulse sequence (18).

NMR data were processed in the format of FTNMR (Hare Research) on a VAX 780 computer using a CSPI Minimap array processor. For the 2D NOE difference spectra, the individual 2D NOE data sets were baseline corrected in ω_1 and ω_2 before and after 2D subtraction.

RESULTS

Figure 2a depicts an ^{15}N-decoupled isotope-edited proton NMR spectrum of an ^{15}N-labeled inhibitor complexed to porcine pepsin (15). As illustrated, excellent suppression of the protons not coupled to ^{15}N labeled nuclei of the

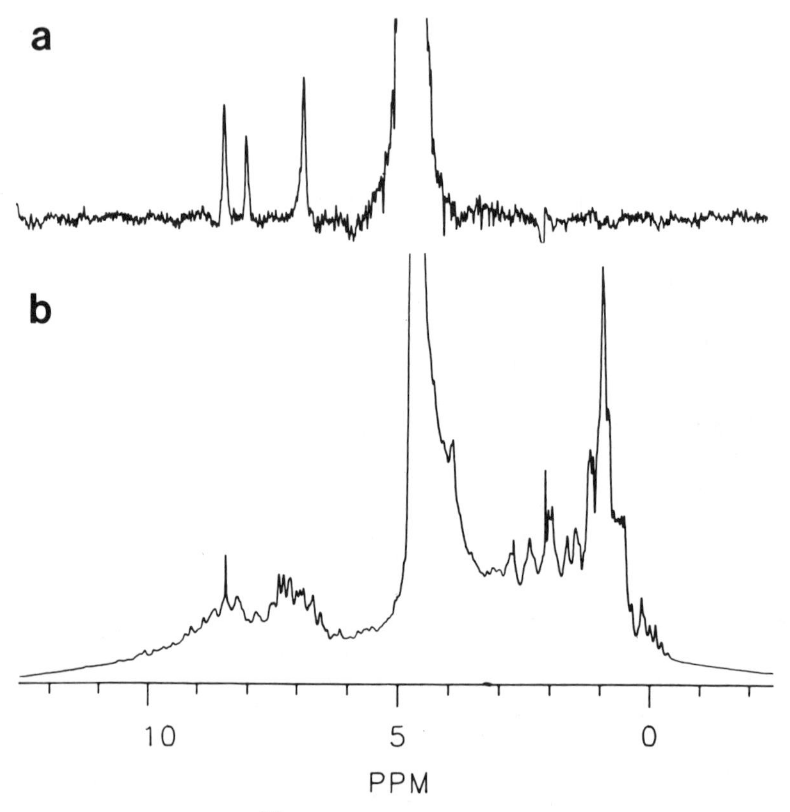

FIGURE 2(a). ^{15}N-decoupled isotope-edited and (b) conventional proton NMR spectra of a H_2O/D_2O (9/1) solution (1 mM) of a pepsin/^{15}N-labeled inhibitor complex (1/1). [Source: S.W. Fesik et. al. (15)].

inhibitor and the enzyme is achieved (compare to the conventional spectrum, Figure 2b). Only three proton NMR signals corresponding to the three amide protons of the inhibitor that are coupled to ^{15}N and the large residual water resonance were observed.

The simplification of the proton NMR spectrum of the pepsin/inhibitor complex allowed the measurement of the NH exchange rates of the bound inhibitor. This was accomplished by measuring the areas of the ^{15}NH signals of the bound inhibitor (assigned by using single ^{15}N-labeled inhibitors) from a series of isotope-edited proton NMR

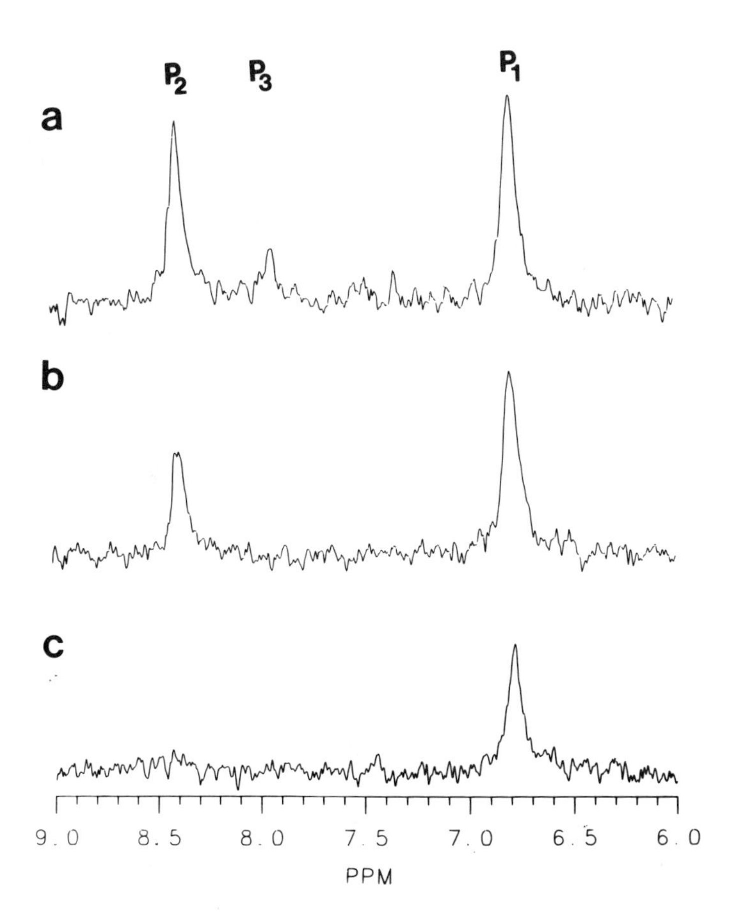

FIGURE 3. Isotope-edited proton NMR spectra (40°C) acquired at a) 20, b) 90, and c) 1200 min. after the preparation of a D_2O solution of the pepsin/inhibitor complex. [Source: S.W. Fesik et. al. (15)].

spectra acquired at various times after the preparation of a ^2H_2O solution of the complex. As shown in Figure 3, the P_3 NH exchanged more rapidly than the P_2 and P_1 NH. These results were interpreted by examining the crystal structures of homologous aspartic proteinase/inhibitor complexes. In all of the crystal structures (19-21), the P_1 NH is completely buried in the active site cleft, consistent with a slow amide exchange rate. The P_2 NH on the other hand, is closer to the outside, and the P_3 NH is nearly at the surface of the enzyme in a location consistent with a more rapid NH exchange rate. Therefore, the amides of the inhibitor that are near the catalytic site exhibit a slower NH exchange rate and are the least exposed to solvent (15).

In Figure 4, a contour map of an isotope-edited two-dimensional NOE experiment of the tri-^{15}N-labeled inhibitor complexed with porcine pepsin is shown. The spectrum is

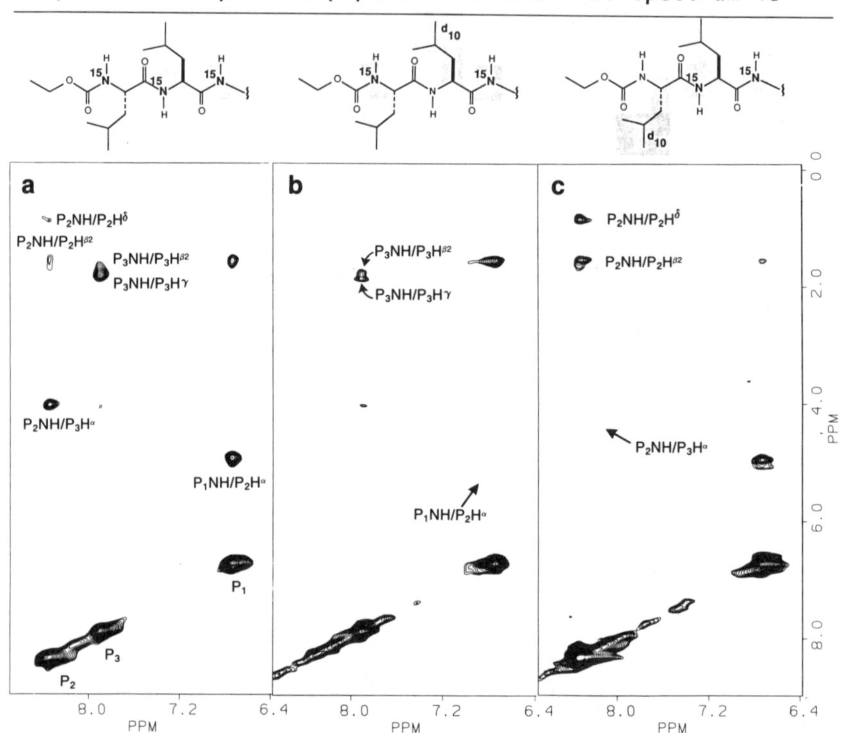

FIGURE 4. Contour plots of ^{15}N-isotope-edited 2D NOE experiments acquired with a mixing time of 50 ms of a pepsin/inhibitor (1/1) complex using the labeled inhibitors shown at the top of the spectra. [Source: S.W. Fesik et. al. (15)].

markedly simplified compared to the unedited 2D NOE spectrum of the complex. Only diagonal peaks corresponding to the labeled P_1, P_2, and P_3 amide protons and NOE cross-peaks between the amide protons and other protons of the enzyme and inhibitor are observed. The NOE cross-peaks were assigned using deuterated inhibitors (16). For example, from the missing cross-peak (Figure 4b, indicated by the arrow) in the NOE spectra obtained with the inhibitor deuterated at the P_2 site and ^{15}N labeled at P_1 and P_3, the NOE observed at this location in Figure 4a could be assigned to an NOE between P_1NH and P_2H^{α}. Similarly, the NOE cross-peak between P_2NH and P_3H^{α} was assigned by comparing the 2D NOE spectra in Figures 4a and 4c.

The NOEs observed between the amide and α protons of adjacent amino acid residues of the enzyme-bound inhibitor are characteristic of an extended main chain conformation (4) in agreement with that found for different inhibitors bound to homologous aspartic proteinases (19-21).

The side-chain conformations of the P_3 and P_2 sites were also determined from isotope-filtered 2D NMR experiments. Figure 5a and 5b depict contour maps of isotope-edited 2D NOE experiments of pepsin/inhibitor

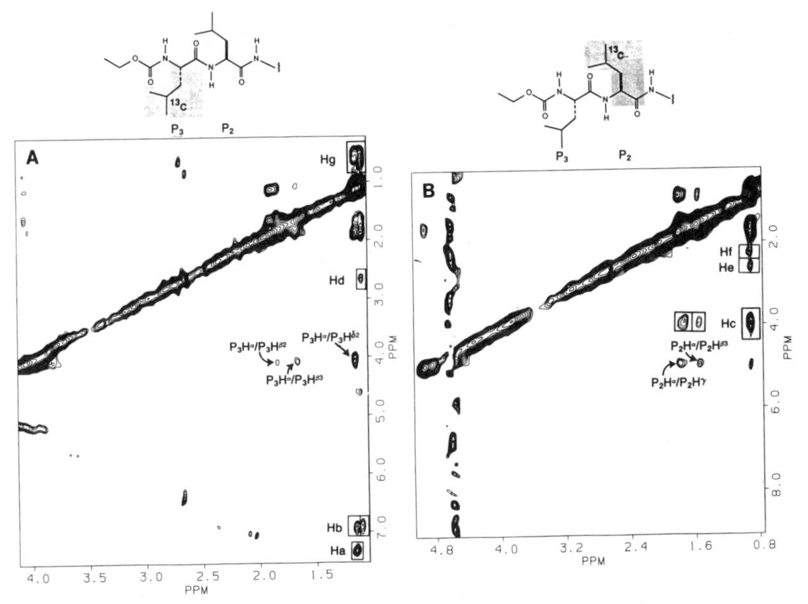

FIGURE 5. Contour plots of ^{13}C-isotope-edited experiments using inhibitors uniformly ^{13}C-labeled (85%) at (A) P_3 and (B) P_2. The boxes indicate NOEs between labeled inhibitor and pepsin. [Source: S.W. Fesik et. al. (15)].

complexes containing the inhibitors uniformly ^{13}C labeled (85%) at the P_3 or P_2 site, respectively. From an analysis of the ^{13}C-isotope-edited 2D NOE data in conjunction with the data obtained from the ^{15}N-isotope-filtered experiments, the χ^1 angle was -60° for the P_2 and P_3 side chains; whereas, χ^2 was found to be ~60° and ~180° for P_2 and P_3, respectively (16).

In addition to the bound conformation of the pepsin inhibitor, information was also obtained on the structure of the active site from NOEs between the inhibitor and pepsin (boxed NOE cross-peaks in Figure 5a and 5b). In order to

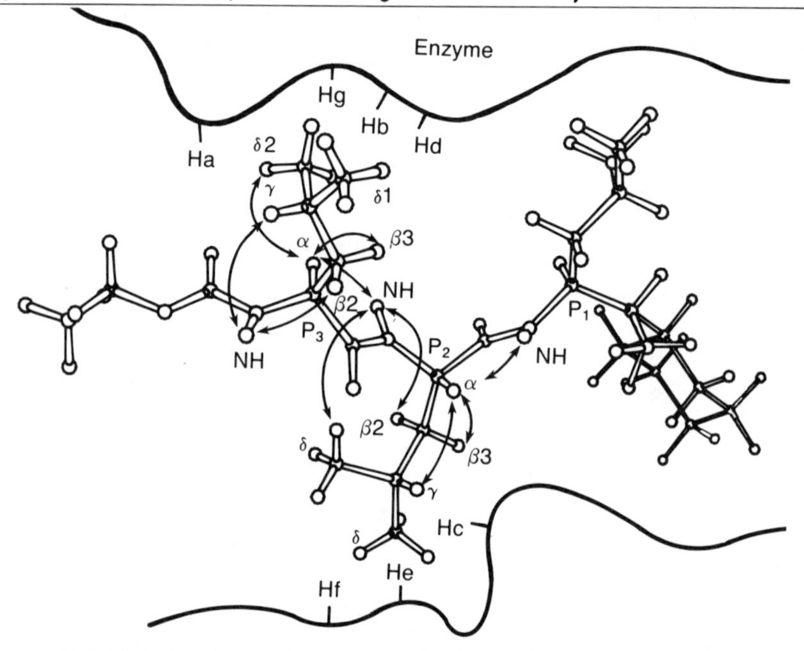

FIGURE 6. Computer-generated model of the bound pepsin inhibitor that is consistent with the NMR data. The arrows indicate NOEs observed between proton pairs of the ligand. Inhibitor/enzyme NOEs are designated by the boxed letters. [Source: S.W. Fesik et. al. (15)].

interpret these inhibitor/enzyme NOEs, a model of this pepsin/inhibitor complex (Figure 6) was built from a partially refined X-ray crystal structure of a similar inhibitor complexed with porcine pepsin. All of the NOE data were rationalized from the model (16).

Another approach for simplifying complicated proton NMR spectra of enzyme/inhibitor complexes is by difference spectroscopy using deuterium labeled ligands. Indeed, by subtracting two-dimensional NOE data sets of pepsin/inhibitor complexes acquired using deuterated inhibitors (at P_3 or P_2) from a 2D NOE spectrum of pepsin complexed with a protonated inhibitor, a marked simplification of the 2D NOE spectrum is obtained. This approach is particularly valuable in cases in which the ligand can be deuterated but not labeled with ^{13}C or ^{15}N. Moreover, this method has a higher sensitivity for large molecules compared to isotope-edited experiments since no fixed delays (1/2J) are required for generating or refocusing antiphase magnetization. The application of this approach for studying enzyme/inhibitor complexes will be described in detail elsewhere.

DISCUSSION

Using isotopically labeled inhibitors and the methods described here, NMR spectroscopy can provide detailed structural information on large enzyme/inhibitor complexes. The conformation of tightly bound ligands can be determined and the ligand's active-site environment can be examined. In contrast to the complete three-dimensional structures that can be obtained by X-ray crystallography, this approach using isotopically labeled ligands only provides structural information about the inhibitor and its immediate vicinity. However, this information may be sufficient for designing more effective inhibitors (22). Moreover, this general approach can be applied to study other biological systems of interest such as ligands bound to soluble receptors and enzyme/substrate interactions.

REFERENCES

1. Goodford PJ (1984). Drug design by the method of receptor fit. J Med Chem 27:557.
2. Stezowski JJ, Chandvasekhar K (1986). X-ray crystallography of drug molecule-macromolecule interactions as an aid to drug design. Ann Reports Med Chem 21:293.

3. Noggle JH, Schirmer RE (1971). "The Nuclear Overhauser Effect. Chemical Applications." New York: Academic Press.
4. Wuthrich K (1986). "NMR of Proteins and Nucleic Acids." New York: Wiley.
5. Clore GM, Gronenborn AM (1987). Determination of three-dimensional structures of proteins in solution by nuclear magnetic resonance spectroscopy. Protein Eng 1:275.
6. Freeman R, Mareci TH, Morris GA (1981). Weak satellite signals in high-resolution NMR spectra: Separating the wheat from the chaff. J Magn Reson 42:341.
7. Bendall MR, Pegg DT, Doddrell DM, Field J (1981). NMR of protons coupled to ^{13}C nuclei only. J Am Chem Soc 103:934.
8. Wilde JA, Bolton PH, Stolowich NJ, Gerlt JA (1986). A method for the observation of selected proton NMR resonances of proteins. J Magn Reson 68:168.
9. Griffey RH, Jarema MA, Kunz S, Rosevear PR, Redfield AG (1985). Isotopic-label-directed observation of the nuclear Overhauser effect in poorly resolved proton NMR spectra. J Am Chem Soc 107:711.
10. Otting G, Senn H, Wagner G, Wüthrich K (1986). Editing of 2D ^{1}H NMR spectra using X half-filters. Combined use with residue-selective ^{15}N labeling of proteins. J Magn Reson 70:500.
11. Bax A, Weiss MA (1987). Simplification of two-dimensional NOE spectra of proteins by ^{13}C labeling. J Magn Reson 71:571.
12. Rance M, Wright PE, Messerle BA, Field LD (1987). Site-selective observation of nuclear Overhauser effects in proteins via isotopic labeling. J Am Chem Soc 109:1591.
13. McIntosh LP, Griffey RH, Muchmore DC, Nielson CP, Redfield AG, Dahlquist FW (1987). Proton NMR measurements of bacteriophage T4 lysozyme aided by ^{15}N isotopic labeling: Structural and dynamic studies of larger proteins. Proc Natl Acad Sci USA 84:1244.
14. Fesik SW, Gampe RT, Rockway TW (1987). Application of isotope-filtered 2D NOE experiments in the conformational analysis of atrial natriuretic factor (7-23). J Magn Reson 74:366.
15. Fesik Sw, Luly JR, Stein HH, BaMaung N (1987). Amide proton exchange rates of a bound pepsin inhibitor determined by isotope-edited proton NMR experiments. Biochem Biophys Res Commun 147:892.

16. Fesik SW, Luly JR, Erickson JW, Abad-Zapatero C (1988). Isotope-edited proton NMR study on the structure of a pepsin/inhibitor complex. Biochemistry 27:8297.
17. Luly JR, Yi N, Soderquist J, Stein H, Cohen J, Perun TJ, Plattner JJ (1987). New inhibitors of human renin that contain novel Leu-Val replacements. J Med Chem 30:1609.
18. Levitt MH, Freeman R, Frenkiel T (1982). Supercycles for broadband heteronuclear decoupling. J Magn Reson 50:157.
19. James MNG, Sielecki A, Salituro F, Rich D, Hofmann T (1982). Conformational flexibility in the active sites of aspartyl proteinases revealed by a pepstatin fragment binding to penicillopepsin. Proc Natl Acad Sci USA 79:6137.
20. Bott R, Subramanian E, Davies DR (1982). Three-dimensional structure of the complex of the Rhizopus chinensis carboxyl proteinase and pepstatin at 2.5-Å resolution. Biochemistry 21:6956.
21. Foundling SI, Cooper J, Watson FE, Cleasby A, Pearl LH, Sibanda BL, Hemmings A, Wood SP, Blundell TL, Valler MJ, Norey CG, Kay J, Boger J, Dunn BM, Leckie BJ, Jones DM, Atrash B, Hallett A, Szelke M (1987). High resolution X-ray analyses of renin inhibitor-aspartic proteinase complexes. Nature 327:349.
22. Fesik SW (1988). Isotope-edited NMR spectroscopy. Nature 332:865.

Frontiers of NMR in Molecular Biology, pages 63-73

SOLUTION NMR STUDIES OF FAB′-PEPTIDE COMPLEXES[1]

P. Tsang, T.M. Fieser, J.M. Ostresh[2], R.A. Houghten, R.A. Lerner and P.E. Wright

Department of Molecular Biology, Research Institute of Scripps Clinic, 10666 North Torrey Pines Road, La Jolla, California 92037

ABSTRACT The results from isotope-edited NMR studies of monoclonal anti-peptide Fab′ complexes are reported here. The complexes consist of Fab′ bound to a series of ^{15}N-labelled twelve residue peptides with the sequence: MHKDFLEKIGGL-NH_2. By labelling the peptide and applying isotope-edited NMR techniques, it has been possible to selectively observe the bound amide proton resonances of the peptide within the complex. Half of the amide proton resonances of the bound peptide have been assigned and studied in this work. The differences among the various NMR parameters of the residues examined are discussed with respect to the antigenic properties of the peptide.

INTRODUCTION

Anti-peptide antibodies are currently of intense interest both because of their importance as research tools and because of their potential use as diagnostics and in medicine (1-6). From a more fundamental point of view, anti-peptide antibodies may be viewed as highly tractable systems for investigation of the classical problems of antibody-antigen recognition and binding. It is still not fully understood how an antibody raised against a flexible peptide fragment of a protein can cross-react with both the

---

[1]This work was sponsored by NIH grants CA27498 and AI19499.
[2]Present address: Multiple Peptide Systems, 10955 John Jay Hopkins Drive, San Diego, CA 92121

peptide immunogen and with its cognate sequence in the folded protein. An adequate description of the mechanism of induction of protein reactive anti-peptide antibodies and of the molecular basis for antigen recognition requires knowledge of the conformational preferences of the immunogenic peptide (both free in solution and coupled to a carrier protein) and information on its structure in the complex with the antibody. In addition, knowledge of the three-dimensional structure of the cognate sequence in the folded protein antigen is important. Using 2D NMR methods, we have previously shown that many immunogenic peptides adopt highly preferred conformations in water solution (7-9) and have postulated a mechanism for induction of protein-reactive anti-peptide antibodies based on the observations (6). In the present paper, we describe NMR approaches to investigate the conformation and dynamics of a peptide bound to an anti-peptide antibody. To make the problem tractable, the Fab′ fragment (molecular weight 55,000) of the antibody is used together with specific isotopic labelling of the peptide.

A major problem inherent to investigations of large molecular weight systems such as an Fab′-peptide complex is the ability to distinguish Fab′ resonances from those of the antigen itself. Due to the recent development in NMR of reverse or inverse-detection techniques (10-17), (*indirect* detection via protons of less abundant, lower sensitivity nuclei such as ^{13}C and ^{15}N, for example), there is an increase in the overall scope of experiments possible for molecular weight systems larger than 15 kDa (18). The technique of isotope-editing is a combined approach involving the specific labelling of the molecule to be studied and the application of reverse-detection NMR methods to observe the protons directly coupled to these labelled nuclei (19-23). In this way, it is possible to observe specifically labelled protons against an overwhelming background of resonances from protons in the macromolecule.

EXPERIMENTAL

Each of the peptides utilized in this work is of sequence: M_1 H_2 K_3 D_4 F_5 L_6 E_7 K_8 I_9 G_{10} G_{11} L_{12}-NH_2 and either singly or doubly labelled with ^{15}N at one of the following amide positions: D_4, F_5, L_6, I_9, G_{10} and G_{11}. These peptides were used to form complexes with the Fab′

fragment of an anti-peptide antibody raised against a peptide comprising residues 69 to 87, the C-helix, of myohemerythrin (24). Reverse-detection methods were then used to observe peptide resonances in the Fab' complex. The protocols used for the synthesis of the peptides, production of the Fab', and the formation of the complex have been described previously (25).

All spectra shown were recorded on a Bruker AM-500 spectrometer at either 298K or 308K at ^{1}H and ^{15}N Larmor frequencies of 499.87 and 50.66 MHz, respectively. Approximately 10,000-40,000 transients were acquired with 8192-32768 points and recycle delays of 1-2 seconds per reverse-detected one-dimensional Fab' spectrum. The NMR sequences and equipment used have been described elsewhere (25).

Typical Fab' NMR samples were pH 5, 0.1M Na deuteroacetate buffer solutions containing 0.5-1 mM Fab'. The stock peptide samples used for the NMR titrations contained approximately 4-6 mM peptide in a 0.1 M Na deuteroacetate buffer, pH 5.

RESULTS AND DISCUSSION

The selective nature of these experiments is apparent by comparing spectra (shown in Fig. 1) recorded from a sample of Fab'-peptide complex using conventional and

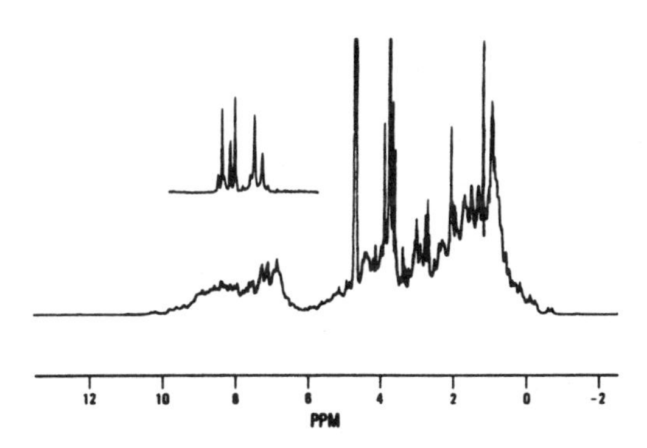

FIGURE 1. Comparison at 308K of the conventional proton spectrum (bottom) of the Fab'-peptide complex and the reverse-detected spectrum (top inset). The sample contains 1 mM Fab' in a pH 5, 0.1 M Na deuteroacetate buffer. The ^{1}H and ^{15}N spectrometer frequencies were 499.87 and 50.66 MHz, respectively. The reverse-detected spectrum was acquired with ^{15}N decoupling.

reverse-detection proton methods. The bottom spectrum in this Figure is the full, conventional proton spectrum. Above the amide/aromatic section of this spectrum, the corresponding region of a ^{15}N-decoupled reverse-detected spectrum is shown. Since the complex contains peptide (labelled with ^{15}N at the two glycine positions) in an approximate molar ratio of 1:2 (Fab′: peptide), resonances from both the bound and unbound peptide should be observed. As expected, four major resonances are observed; other resonances in the spectrum may be attributable to impurities in the peptide sample, as well as the likely occurrence of less specific binding of the peptide to the Fab′ in the presence of excess peptide. From this comparison the benefits of editing are obvious, allowing specific observation of the resonances from protons bound to the labelled amide residues of the peptide only.

In Figure 2, NMR spectra acquired using reverse techniques at different points of a titration of Fab′ with a peptide labelled at the two glycine residues, G_{10} and G_{11}, are shown. Aside from a broad, low intensity baseline hump

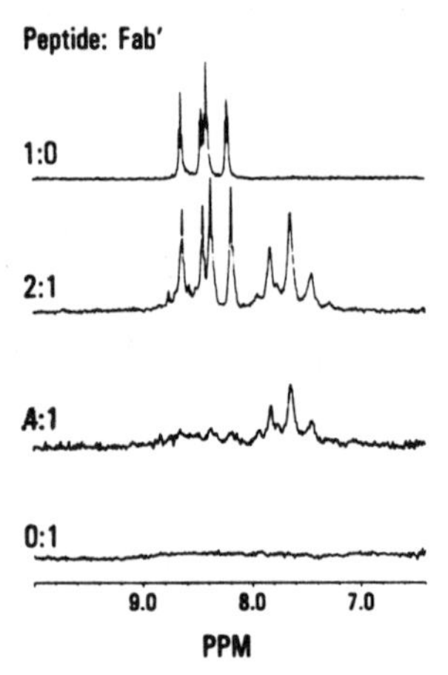

FIGURE 2. Reverse-detected spectra acquired without ^{15}N decoupling from various samples of Fab′ and peptide (labelled at G_{10} and G_{11}), ranging from Fab′ only (0:1 peptide:Fab′) to peptide only (1:0 peptide:Fab′). Intermediate ratios of peptide to Fab′ corresponding to 0.4:1 and 2:1 peptide: Fab′ were acquired at 308K at 499.87 (^{1}H) and 50.66 (^{15}N) frequencies. All spectra were recorded identically, using the same buffer, pH, temperature and spectrometer conditions to minimize differences among the resonances of the bound and unbound forms of the peptide.

attributable to signals from Fab' amide protons attached to natural abundance ^{15}N nuclei, major resonance intensity is detected at the beginning of the titration (corresponding to the spectrum at 0:1 peptide:Fab').

As increasing concentrations of peptide are titrated into the antibody, a set of resonances appears first at 7.51-7.72 ppm followed by a second set around 8.20-8.44 ppm when the peptide is in excess. From the close correspondence between the chemical shifts of the second set of resonances and those of the free peptide (upper spectrum), these resonances are attributed to unbound peptides. The first group of resonances is assigned to the bound form of the peptide. Since the ^{15}N-edited ^{1}H NMR spectrum recorded from a sample of Fab' complexed with doubly ^{15}N labelled peptide contains two sets of overlapping doublet resonances (resembling a triplet pattern), it appears that the peptide binds to one major site. From the observation of resolvable, distinct resonances from the free and bound forms of the peptide, the chemical exchange rate of the peptide is slow on the chemical shift timescale. More specifically, the exchange rate estimated from the smallest observed chemical shift differences between the bound and unbound resonances in both the ^{15}N and ^{1}H spectra is less than approximately 90 sec^{-1}.

The "bound" resonances clearly arise from a very specific binding interaction between the Fab' and the peptide. However, there is also some evidence for less specific binding of the peptide to the Fab'. Figure 3 shows the reverse-detected spectra (obtained without ^{15}N-decoupling) from two Fab'-peptide complexes in the presence of excess peptide. The two peptides in both cases are doubly labelled. The first of these, G_{10} G_{11}, has been discussed above; the second is labelled at the D_4 F_5 residues. The sample containing the complex with the latter peptide contains an approximate ratio of 3:1 peptide:Fab' versus 2:1 for the other complex shown in this Figure. In the D_4 F_5 peptide conplex, the bound resonances have been attributed to the broad doublet resonances at around 9.6 and 8.9 ppm; the unbound peptide is observed at around 8.38-8.10. In both cases, in addition to the resonances of the bound and unbound peptide, some lower intensity, generally sharper resonances can also be seen. Neither these resonances nor those of the free (unbound) peptide are observable following equilibrium dialysis of the sample against buffer. Their lower intensities and significantly different appearances relative to resonances of either the bound or free forms implies that they arise from much less specific and lower affinity interactions of the peptide with the Fab'.

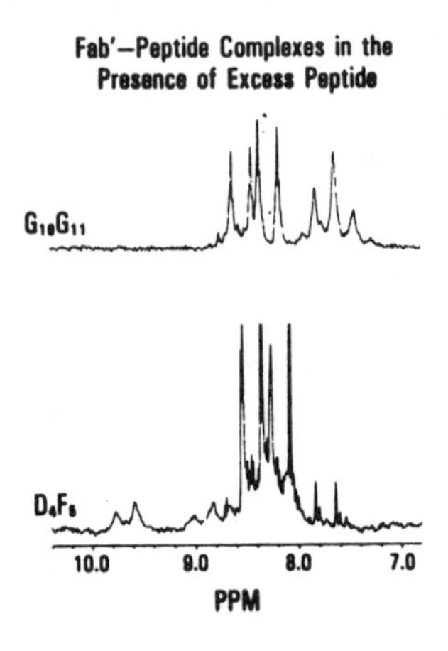

FIGURE 3. ^{15}N-undecoupled reverse-detected spectra acquired from two samples containing complexes in the presence of excess peptide. The top spectrum was obtained from a sample containing Fab' complexed with peptide labelled at G_{10} and G_{11} in an approximate ratio of 2:1 peptide:Fab'. The bottom spectrum was obtained from a complex containing D_4 F_5 labelled peptide which is present in a 3:1 peptide:Fab'ratio. The samples were run at 499.87 and 50.66 MHz and 298K.

NMR titrations similar to the one described earlier were conducted using other labelled peptides. These peptides included the doubly labelled D_4 F_5 peptide, as well as peptides singly labelled with ^{15}N at the F_5, L_6, I_9 and G_{10} residues. In all cases, the same basic trend was observed; a set of resonances corresponding to the bound peptide was observed first, followed by the appearance of a second set of resonances once the peptide was in excess at a frequency corresponding to that of free peptide. The spectra recorded from each of these complexes are shown in Figure 4. In all cases except one (L_6), the samples contained peptide in equimolar or lesser amounts.

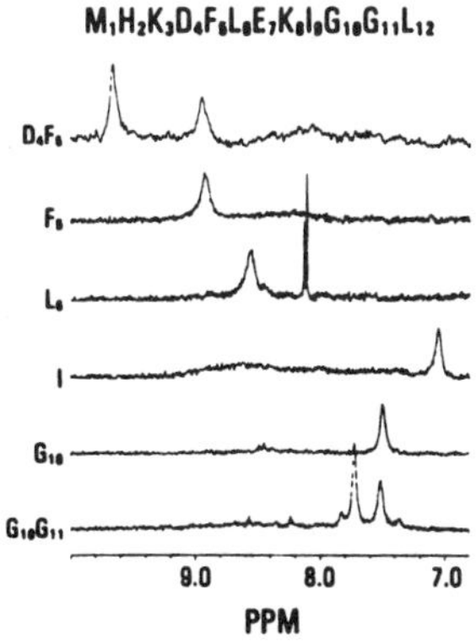

FIGURE 4. The resonance assignments of the six peptide residues are shown here. The labelled peptides used in the complexes are indicated at the left. The assignment of the doubly labelled residues may be deduced by comparison of the doubly and singly labelled peptide complex spectra. The D,F and G,G resonances were assigned this way. All spectra were recorded via reverse detection with ^{15}N-decoupling at 499.87 (^{1}H) and 50.66 (^{15}N) MHz and 308 K.

A diagram summarizing the results obtained from all the labelled peptide complexes studied is shown in Figure 5. The solid and diagonally striped bars represent the free and bound peptide resonances studied thus far. The relative widths and positions of these bars represent the linewidths and chemical shifts, respectively, of the observed resonances. From this figure, the trends in the two most easily observed properties, i.e., frequency shifts and linewidth changes upon binding of the peptide to the Fab', may be followed. The residues D, F, L and I give rise to bound resonances whose linewidths all appear to be approximately twice the linewidth of the narrowest resonance observed from all of the complexes, i.e., G_{11} (18 Hz). The resonance of G_{10} is slightly broader (24 Hz). The resonance linewidths of the remaining F, L, I and D sites are in the range of 30-47 Hz. The observed range of ^{1}H chemical shift changes upon binding is about 0.4-1.3ppm.

Immunological studies have assigned the epitope to residues DFLEKI (24), a region which adopts nascent helical structures in the myohemerythrin C-helix peptide in aqueous solution (9). In light of this information, the NMR

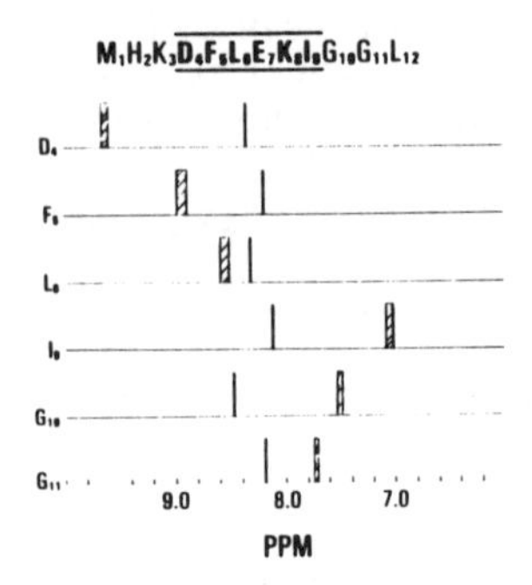

FIGURE 5. Schematic representation of the various bound and unbound resonances observed from the different complexes. The diagonally striped bars are those of the bound and the solid bars are of the free amide proton resonances. The chemical shifts and linewidths of these resonances are indicated by the relative positions along the axis and the overall widths of these bars, respectively. The residues in the epitope region of the peptide are shown in boldface in the sequence shown at the top of the figure.

results obtained for several of the residues may be more easily interpreted. Assuming that other line broadening effects are not significant (such as differential solvent exchange or chemical exchange, for example), the linewidth differences among the various resonances may reflect differential residue mobility. The maximum linewidth difference seen here is more than twofold, 18 Hz for G_{11} compared to 47 Hz for F_5. This is consistent with a model in which residues in the epitope are considerably immobilized upon binding to the antibody, while residues outside the epitope are much less so. With the exception of residue D_4, a mobility gradient exists, based upon the gradual increase in linewidth as one moves from G_{11} towards the N-terminal residues. In this respect, the behavior of the peptide is similar to N- and C-terminal segments of intact proteins since the residues near the end often exhibit greater mobility than those present in more rigid, folded sections of the protein. The estimated average ^{14}N-coupled amide proton resonance linewidth appears to be 40-50Hz for the Fab'. Line-broadening effects due to the attached ^{14}N quadrupolar nucleus (26) are not significant for systems in the molecular weight range of the Fab', i.e., 55 kDa. The maximum linewidths of the amide proton resonances of the bound peptide are comparable to those of the Fab'. The broader resonances of the bound peptide arise from residues within the epitope. The few residues outside the epitope that have been labelled appear to be considerably more mobile.

The present experiments show that we are able to monitor interactions between individual residues of the peptide antigen and the Fab' using isotope-edited NMR methods. Specific properties of the individual sites of the peptide, such as mobility, for example, may be probed by conducting relaxation studies on the labelled peptide complexes. More detailed NMR investigations, including measurement of NOEs, are being undertaken to investigate the conformation of the antibody-bound peptide. It is clear that isotope-edited NMR experiments hold great potential for detailed studies of peptide-antibody and peptide-receptor interactions.

ACKNOWLEDGEMENTS

The authors would like to thank Dr. M. Rance for helpful suggestions and discussions, and L. Tennant and R. T. Samodal for their expert technical aid and assistance throughout this work.

REFERENCES

1. Lerner RA (1982). Tapping the immunological repertoire to produce antibodies of predetermined specificity. Nature 299:592.
2. Lerner RA (1984). Antibodies of predetermined specificity in biology and medicine. Adv Immunol 36:1.
3. Steward MW, Howard CR (1987). Synthetic peptides: a next generation of vaccines? Immunol Today 8:51.
4. Porter R, Whelan J (eds) (1986). Synthetic peptides as antigens. CIBA Symp. Vol. 119, p 307.
5. Arnon R (1986). Synthetic peptides as the basis for future vaccines. Trends Biochem Sci 11:521.
6. Dyson HJ, Lerner RA, Wright PE (1988). The physical basis for induction of protein-reactive antipeptide antibodies. Ann Rev Biophys Biophys Chem 17:305.
7. Dyson HJ, Cross KJ, Houghten RA, Wilson IA, Wright PE, Lerner RA (1985). The immunodominant site of a synthetic immunogen has a conformational preference in water for a type II reverse turn. Nature 318:480.
8. Dyson HJ, Rance M, Houghten RA, Lerner RA, Wright PE

(1988). Folding of peptide fragments of proteins in water solution. 1. Sequence requirements for the formation of a reverse turn. J Mol Biol 201:161.

9. Dyson HJ, Rance M, Houghten RA, Wright PE, Lerner RA (1988). Folding of peptide fragments of proteins in water solution. 2. The nascent helix. J Mol Biol 201:201.
10. Vidusek DA, Roberts MF, Bodenhausen G (1982). Indirect detection of ^{199}Hg NMR to characterize adducts of ethylmercury phosphate with amino acids and ribonuclease. J Amer Chem Soc 104:5452.
11. Muller L (1979). Sensitivity enhanced detection of weak nuclei using heteronuclear multiple quantum coherence. J Amer Chem Soc 101:4481.
12. Maudsley AA, Muller L, Ernst RR (1977). Cross-correlation of spin-decoupled NMR spectra by heteronuclear two-dimensional spectroscopy. J Magn Reson 29:463.
13. Redfield AG (1983). Stimulated echo NMR spectra and their use for heteronuclear two-dimensional shift correlation. Chem Phys Lett 96:537.
14. Wagner G, Brühwiler D (1986). Toward the complete assignment of the α-carbon nuclear magnetic resonance spectrum of the basic pancreatic trypsin inhibitor. Biochem 25:5839.
15. Live DH, Davis DG, Agosta WC, Cowburn D (1984). Observation of 1000-fold enhancement of ^{15}N NMR via proton-detected multi-quantum coherences: studies of large peptides. J Amer Chem Soc 106:6104.
16. Griffey RH, Redfield AG (1987). Proton-detected heteronuclear edited and correlated nuclear magnetic resonance and nuclear Overhauser effect in solution. Quart Rev Biophys 19:51.
17. Bax A, Griffey RH, Hawkins BL (1983). Correlation of proton and nitrogen-15 chemical shifts by multiple quantum NMR. J Magn Reson 55:301.
18. Wüthrich K (1986). NMR of proteins and nucleic acids. John Wiley and Sons, New York, p 259.
19. Ortiz-Polo G, Krishnamoorthi R, Markley JL, Live DH, Davis DG, Cowburn D (1986). Natural-abundance ^{15}N NMR studies of turkey ovomucoid third domain. Assignment of peptide ^{15}N resonances to the residues at the reactive site region via proton-detected multiple quantum coherence. J Magn Reson 68:303.

20. Rance M, Wright PE, Messerle BA, Field LD (1987). Site-selective observation of nuclear Overhauser effects in proteins via isotope labelling. J Amer Chem Soc 109:1591.
21. McIntosh LP, Dahlquist FW, Redfield AG (1987). Proton NMR and NOE structural and dynamic studies of larger proteins and nucleic acids aided by isotope labels: T4 lysozyme. J Biomolec Struc Dyn 5:12.
22. Fesik SW (1988). Isotope-edited NMR spectroscopy. Nature 332:865.
23. Fesik SW, Luly JR, Stein HH, BaMaung N (1987). Amide proton exchange rates of a bound pepsin inhibitor determined by isotope- edited proton NMR experiments. Biochem Biophys Res Comm 147:892.
24. Fieser TM, Tainer JA, Geysen HM, Houghten RA, Lerner RA (1987). Influence of protein flexibility and peptide conformation on reactivity of monoclonal antibodies with a protein α-helix. Proc Nat Acad Sci USA 84:8568.
25. Tsang P, Fieser TM, Ostresh JM, Lerner RA, Wright PE (1988). Isotope-edited NMR studies of Fab'-peptide complexes. Pept Res 1:87.
26. Llinas M, Klein MP, Wüthrich K (1978). Amide proton spin-lattice relaxation in polypeptides. Biophys J 24:849.

Frontiers of NMR in Molecular Biology, pages 75-87

STUDIES OF THE 3D STRUCTURE OF COMPLEMENT PROTEIN C5A AND C5A MUTANTS BY 2D AND 3D NMR

Erik R.P. Zuiderweg, David G. Nettesheim, Stephen W. Fesik, Edward T. Olejniczak, Wlodek Mandecki[+], Karl W. Mollison, Jonathan Greer and George W. Carter

Pharmaceutical Discovery and [+]Corporate Molecular Biology, Abbott Laboratories, Abbott Park, Illinois 60064, U.S.A.

The tertiary structure for the complement protein C5a in solution was calculated from NMR data. It was demonstrated that contiguous regions of slowly exchanging amide protons identify helical regions in this protein. It was found that the correlation time of the C-terminal residues of C5a is much shorter than in the remainder of the molecule from quantitative NOESY/ROESY measurements. Additional distance constraints were defined from a heteronuclear three dimensional NMR experiment on 0.7 mM N-15 labeled protein. Mutant C5a species were investigated with 2D NMR. It could be demonstrated that mutants R40 > G and I41 > M produced their deleterious effect on C5a affinity by very local conformational changes, if any; mutant A26 > M causes its deleterious effect by a long range effect centered on the loop between helices I and II of the protein.

INTRODUCTION

The 74 amino acid C5a protein is the principal inflammatory molecule derived from the complement system. This potent mediator causes vasodilation, increases vascular permeability, induces the contraction of smooth muscle and triggers histamine release from mast cells. In addition, the protein stimulates the locomotion (chemokinesis) and recruitment (chemotaxis) of polymorphonuclear leukocytes to sites of inflammation and triggers these cells to release tissue digesting enzymes and other damaging substances (for a review, see Ref. 1). Because of this broad range of biological properties, C5a has been implicated as a causative or aggravating agent in a variety of inflammatory and allergic diseases. Compounds which inhibit the inflammatory actions of C5a should therefore be useful therapeutic agents; the current structural NMR studies on this protein and its mutants are aimed towards the design of such compounds. Prior to these studies, only a model structure was available for the protein (2,3).

Although the interaction of C5a with its leukocyte receptors has been studied extensively by several groups, it is still poorly understood; virtually all parts of the molecule have been implicated in receptor binding. The results of the joint biological assay/NMR structure mutagenesis study of C5a presented here indicate the existence of two distinct receptor binding sites on the molecule.

EXPERIMENTAL PROCEDURES

C5a Structure.

The NMR data were obtained from a solution of 7 mM C5a, pH 2.3 at 10 C, recorded at 500 MHz on GE GN500 and Bruker AM500 spectrometers.
The calculations of C5a structures were carried out on a VAX 8600 computer using DISMAN (4) followed by XPLOR (5). Fourteen different starting structures were generated in which the back bone conformations were set in a random extended structures. DISMAN calculations were carried out for each of these fourteen structures using a distance contraints list containing NOEs only (285 entries) resulting in 9 equally well converged structures and calculations were carried out in which the NOE list was augmented with hydrogen bond and disulfide constraints (346 entries) resulting in 11 equally well converged structures.
The average RMSD in the latter ensemble of structures is 1.67 A and 2.13 A for all C-alpha atoms and all heavy atoms respectively. The eleven structures in this ensemble were subjected to a 1.25 ps restrained molecular dynamics minimization (XPLOR) using a constraints list augmented with 297 mimimum approach constraints which reflected the unambiguous lack of NOEs between protons in the NOESY spectrum. The average RMSD values did not change upon MD minimization; all constraints were fulfilled within 0.6 A for all structures.

Three dimensional NMR.

The heteronuclear three dimensional (6) NOESY-HMQC spectrum was obtained from 0.7 mM uniformly N-15 labeled C5a in water, pH 2.5 and 20 C. The pulse sequence used was:

```
Protons:  90 - t1 - 90 - Tm - 90 - T -     180     - T - t3
Nitrogen:                                90 - t2 - 90
```

The data were collected as a set of 37 complex (t2) 2D spectra composed of 100 complex t1 values and 1024 complex t3 datapoints. Each acquisition consisted of 16 scans preceeded by 4 dummy scans. Total actual instrument time for the 3D experiment was 177 hours of which 47 hours were required for I/O operations. The data were processed on a CSPI mimimap array processor interfaced to a Vax 8350 computer using software written at Abbott Labs for a total transformed data size of 1024x128x384 points.

Mutagenesis.

Mutant C5a protein was obtained (7-9) from synthetic genes constructed using restriction fragment replacement (10) or oligonucleotide-directed double-strand break repair (11). NOESY spectra were collected from 0.8 mg quantities of the species Arg 40 > Lys, Gly; Ile 41 > Met and Ala 26 > Met for 0.2 mM solutions. The conditions were chosen to allow for comparison with spectra of native C5a (pH 2.35, 10 C). The NOESY data (200 ms mixing) in water (Ala 26 > Met) were recorded in 90 hours.

RESULTS AND DISCUSSION

Tertiary Structure of C5a.

The tertiary structure of C5a in solution (Figure 1) was determined from a large number of distance constraints using the dihedral angle folding program DISMAN (4) followed by restrained molecular dynamics (12,13).

FIGURE 1. An ensemble of structures of C5a 1-63, obtained from DISMAN and Molecular Dynamics calculations.

Short and medium range constraints. Based on the assignments of many proton resonances in the C5a NMR spectrum (14), we have identified for this protein a large number of inter-residue nuclear Overhauser effects (NOEs) in the two dimensional NOE spectrum. Figure 2 shows the short and medium range NOEs observed in C5a. It is apparent that C5a is an

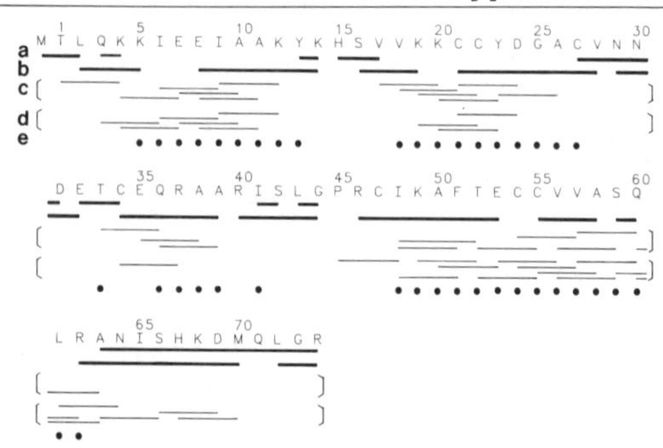

FIGURE 2. Sequential (a,b) and Medium Range (c,d) NOEs, and slowly exchanging amide hydrogens (e). (23)

alpha helical protein; long stretches of sequential amide-amide proton connectivities (lane b), together with intermediate range NOEs (lanes c and d) and the absence of (strong) sequential amide-alpha NOEs (lane a) are seen to accumulate in four separate contiguous regions of the protein: 4-12, 18-26, 34-39 and 46-63. Distances for many of the sequential NOEs were determined from build-up studies using NOESY datasets of 25, 50, 75 and 100 ms mixing time (15).

Dynamics of the C5a backbone. A simultaneous presence of strong amide-alpha and amide-amide NOEs was observed for residues beyond Ile 65 (Figure 2) and semi quantitative NOE studies showed that corresponding distances are apparently approximately equal and about 3 A, a conformation which is very unfavorable. A study of the dynamics of the C5a backbone offers an explanation for this observation. Following the suggestion of Davis (16), rotational correlation times of internuclear backbone vectors were obtained from the ratio of build-up rates of the longitudinal Overhauser effect crosspeaks (NOEs) and rotating-frame Overhauser effect cross peaks (ROEs):

$$\frac{\sigma_{NOE}}{\sigma_{ROE}} = \frac{\frac{6}{1+4\omega^2\tau_c^2} - 1}{\frac{3}{1+\omega^2\tau_c^2} + 2}$$

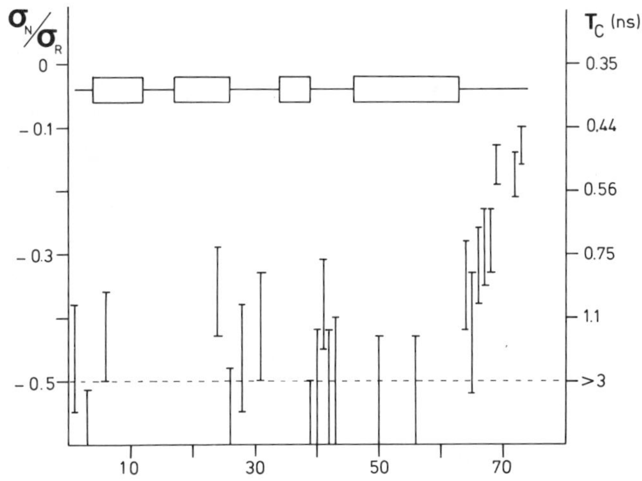

FIGURE 3. Ratio of NOE and ROE build-up rates versus the C5a sequence; the corresponding correlation times at 500 MHz are given at the right.

The experimental ROE build-up rates were corrected for r.f. offset effects. The resulting correlation times are shown in Figure 3, from which it is seen that the C-terminus of C5a is far more mobile than the remainder of the molecule, thus accounting for the observed averaging of alpha amide and amide-amide NOEs in this region. These results are important because the C-terminus of C5a contains one of the receptor binding sites (see below). As a consequence, no structure determination of the C-terminus of C5a was attempted; all structures comprise residues 1 - 63 of C5a only.

Hydrogen Bonding. In the helical segments, long contiguous stretches of amide protons which exchange order of magnitude slower than amide protons in unstructured peptides are observed (Figure 2e), most likely reflecting hydrogen bonding, rather than solvent inaccessibility of the amide moieties in a small protein like C5a. This was rigorously tested by comparing an ensemble of structures generated with additional constraints from an alpha-helical hydrogen bonding pattern for the slowly exchanging amides within the helices with an ensemble calculated from NOE constraints only. It was found that these additional constraints could be incorporated in the calculations, without increasing the residual constraints violations of the resulting structures. Furthermore, as shown in Figure 4 where the ensembles of structures for both calculations are

FIGURE 4. Ensembles of C5a structures computed without (dashed) and with (drawn) hydrogen bond constraints. (23)

superimposed, the ensemble of hydrogen bond constrained structures falls within the ensemble of structures which were not subject to these constraints. It is found that the helical pitch is the same in both ensembles and that slow exchange of long, contiguous regions of amide protons in this molecule is primarily caused by hydrogen bonding rather than by solvent inaccessibility.

Long-Range Constraints. NOE cross peaks were analyzed in a NOESY spectrum recorded under conditions of spin diffusion (200 ms mixing time). Once it was determined which residues were spatially close from the spin diffusive data, the cross peaks still observable in a non spin-diffusive spectrum (recorded with a NOE mixing time of 50 ms) were assigned to the specific proton pairs within those residues. Together with the short and medium range NOEs, a data set of 285 interresidue NOE distance constraints could be established for the region 1-63 of C5a from NOESY datasets recorded with a 50 ms mixing time.

Additional Constraints from Heteronuclear 3D NMR. Even for a relatively small protein like C5a, considerable resonance overlap limits the number of NOEs obtainable from 2D NOESY spectra. We used heteronuclear three dimensional NMR spectroscopy (6) on C5a to probe if additional constraints can be obtained from this technique. A 3D NOESY-HMQC spectrum was recorded on a 0.7 mM solution of C5a, uniformly enriched with N-15. The experiment resolves overlapping NOESY cross peaks by editing 2D NOESY with respect to the amide nitrogen chemical shift in a third dimension, Figure 5. NOESY resonance overlap is effectively resolved with the

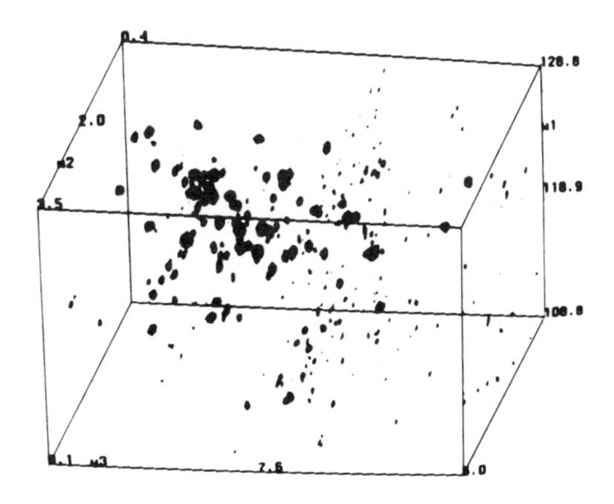

FIGURE 5. The amide proton (w3), methylene-methyl (w2), amide nitrogen (w1) region of a three dimensional heteronuclear NMR spectrum of 0.7 mM C5a.

technique, allowing sequential assignments to be made more reliably and additional distance constraints such as NOEs between amide protons and sidechain protons which are unaccessible in the 2D data, to be obtained.

Comparison with the C3a structure. The conformation of C5a is closely related to that of C3a (2,3,17) for which a crystal structure has been determined (18). In Figure 6 we

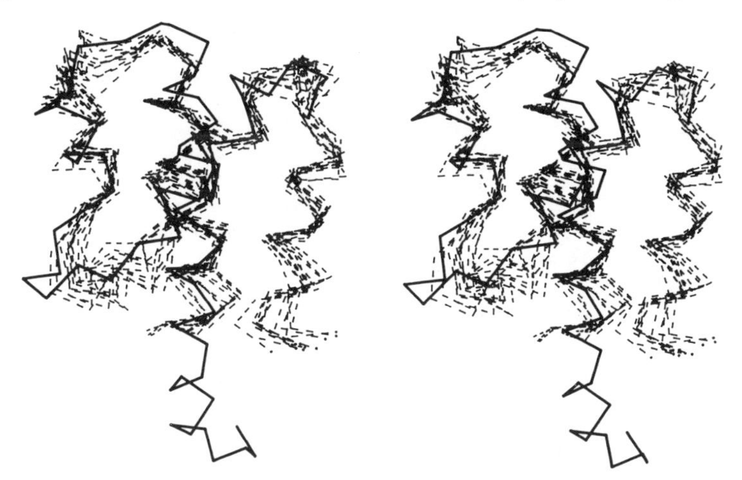

FIGURE 6. Superimposition of the C5a NMR ensemble with the crystal sructure of C3a (drawn) (23).

have superimposed the backbone of C3a with the ensemble of C5a structures. It is found that the locations of the helices in C5a correspond very closely to those in C3a. Apparently, large differences between the structures exist for the terminal regions. In C5a in solution, the C-terminus is only poorly defined, where for C3a in the crystal a regular helix is found followed by a peptide in a stable beta-turn type conformation. In C5a in solution a docked

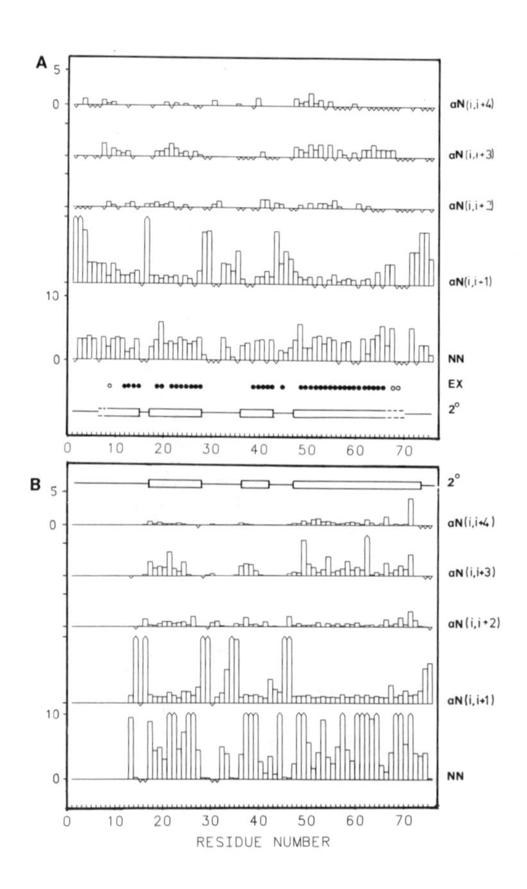

FIGURE 7. Comparison of short and medium range NOEs and hydrogen exchange as observed in C3a in solution (top) with computed NOE data from crystal coordinates (bottom). The lanes labeled "2°" indicate the deduced secondary structures with boxes for helices.

N-terminal helix is observed as proposed (2,3) which corresponds to a disordered region in the crystal for C3a. However, a NMR investigation of the secondary structure of C3a (19) showed that in solution disorder occurs for the C-terminus and that helicity prevails for the N-terminus (Figure 7) just as observed for C5a. The difference between C3a in solution and C3a in the crystal is caused by the crystal packing; the C-termini of C3a molecule pairs interact in the crystal (18) stabilizing the C-terminal helix while the packing does not leave space for a helical N-terminus, docked to the core of the molecule (2,3). Thus, the differences in biological activities between C3a and C5a (C3a does not display chemokinesis nor chemotaxis activity) are very unlikely to be caused by major conformational differences between the two molecules. Rather, the differentiation in biological activities is due to the differences in the chemical characteristics of their surface residues, which are very different indeed (2,3).

C5a Mutagenesis.

Regions of C5a that interact with the receptor may be well dispersed in the protein. The C-terminus of the molecule is of importance for its biological activity as follows from the fact that removal of the C-terminal arginyl residue results in a large decrease in receptor binding affinity (20). However, upon complete removal of 11 C-terminal residues, C5a (1-63) retains a significant fraction of its receptor-binding affinity (10E-6 M versus 10E-11 M for native C5a, Ref. 21). A NMR study of this C5a derivative showed that although the removal of the 11 C-terminal residues did negatively affect the long-term stability of the protein, no significant chemical shift changes were detected for the remaining residues in the 2D spectra of fresh solutions of this derivative, except for those close to the clipping site (Nettesheim and Zuiderweg, unpublished data). Since no conformational changes occurred in the core of the molecule it is therefore indicated that C5a has at least two receptor binding sites, one in the core and one at the C-terminus, and that these sites are independent of each other.

In order to identify the residues involved in the C5a receptor interface, a large number of C5a mutant proteins were prepared and assayed. Mutations of residues in the C-terminal region of C5a caused activity changes (22) as expected.

Several mutations in the core region also affected the activity; an Arg > Gly mutation at residue 40 and an Ile > Met change at residue 41 both caused a decrease in potency. Both species were investigated with NOESY data; by comparing the chemical shifts and the NOEs, it was found that these species retained their native conformation and that only small changes occurred in the immediate vicinity of the mutated residues. Thus, these mutations suggest an area of a receptor interface centered around Arg 40.

Interesting results were obtained for the mutant Ala 26 > Met. Although this residue is totally buried, its mutation negatively affected C5a receptor affinity. The NMR investigation of this molecule revealed large changes in chemical shifts (Fig. 8) which necessitated a reassignment of the spectrum. It was found that residues already known to be located in a C5a receptor interface (e.g. C-terminal residues and residues in the area of Arg 40) did not change the either their amide or alpha proton shift. In contrast, residues in the loop region between the first and second helices show large shift changes while a region around Val 56 is affected to much lesser extent; the latter region is spatially close to the mutagenesis site. In addition, multiple conformations were observed for residue 10 and the NOE Ala 10 - Tyr 23 disappeared for the predominant conformation.

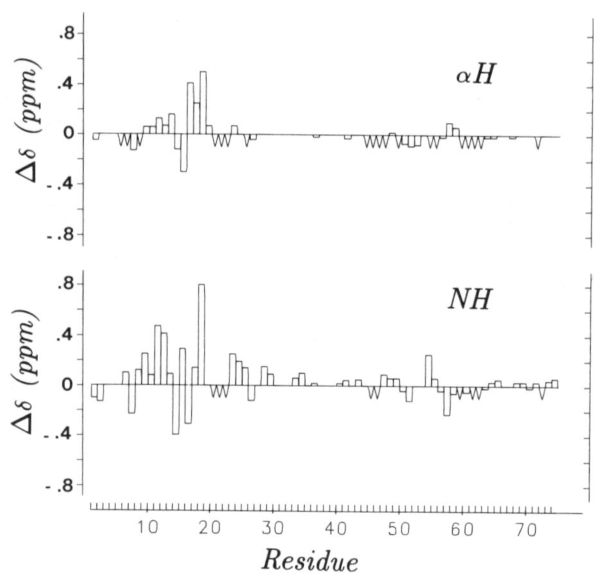

FIGURE 8. Chemical shift changes in mutant C5a Ala 26 > Met with respect to native C5a.

The NMR data indicate therefore that a destabilization of the helix I - core interface occurs in this mutant. Other mutagenesis studies revealed no involvement of the N-terminal residues in receptor binding (22). Therefore, the current results suggest that a conformational change in the loop region is associated with the change in receptor affinity. The core receptor binding site is therefore likely located at the "top" of the molecule in the orientation of Figure 1.

Interestingly, in C3a, residue 26 is a methionine, just as in the mutant, and the N-terminal helix of C3a is not docked in a very stable way either. It may therefore be due to the character of residue 26 in C3a, that its N-terminal helix could be displaced so easily in the crystal, giving rise to the observed discrepancy between crystal and solution structure for this protein.

ACKNOWLEDGEMENTS

We thank Ms. M.A. Shallcross, Ms. M. Seavy-Cork, Dr. J. Henkin, Mr. R.P. Edalji and Mr. R.T. Gampe, Jr. for their contributions to the different aspects of C5a protein preparation.

REFERENCES

1. Hugli TE (1981). The structural basis for anaphylatoxin and chemotactic functions of C3a, C4a, and C5a. Crit. Rev. Immunol. 1:321.
2. Greer J (1985). Model structure for the inflammatory protein C5a. Science 228:1055.
3. Greer J (1986). Comparative structural anatomy of the complement anaphylatoxin proteins C3a, C4a and C5a. Enzyme 36:150.
4. Braun W, Go N (1985). Calculation of protein conformations by proton-proton distance constraints. J Mol Biol 186:611.
5. Brunger AT (1987). XPLOR, Yale University and Howard Hughes Medical Institute, New Haven CT.
6. Fesik SW, Zuiderweg ERP (1988). Heteronuclear three-dimensional NMR spectroscopy: a strategy for the simplification of homonuclear two-dimensional NMR spectra. J Magn Reson 78:588.
7. Mandecki W, Powell BS, Mollison KW, Carter GW, Fox JL (1986). High-level expression of a gene encoding the human complement factor C5a in Escherichia coli. Gene 43:131.

8. Mollison KW, Fey TA, Krause RA, Mandecki W, Fox JL, Carter GW (1987). High-level gene expression and recovery of recombinant human C5a from Escherichia coli. Agents and Actions 21:366.
9. Carter GW, Mollison KW, Fayer L, Fey TA, Krause R, Henkin J, Edalji RP (1986). Purification and characterization of recombinant human C5a. Complement 2:15.
10. Mandecki W, Mollison KW, Bolling TJ, Powell BS, Carter GW, Fox JL (1985). Chemical sythesis of a gene encoding the human complement fragment C5a and its expression in Escherichia coli. Proc Natl Acad Sci USA 82:3543.
11. Mandecki W (1986). Oligonucleotide-directed double-strand break repair in plasmids of Escherichia coli: a method for site-specific mutagenesis. Proc Natl Acad Sci USA 83:7177.
12. Kaptein R, Zuiderweg ERP, Scheek R, Boelens R, Van Gunsteren WF (1985). A protein structure from nuclear magnetic resonance data: lac repressor headpiece. J Mol Biol 182:179.
13. Clore GM, Gronenborn AM, Brunger AT, Karplus M (1985). Solution conformation of a heptadecapeptide comprising the DNA binding helix F of the CAP protein of Escherichia coli: combined use of 1H NMR and restrained molecular dynamics. J Mol Biol 186:435.
14. Zuiderweg ERP, Mollison KW, Henkin J, Carter GW (1988). Sequence-specific assignments in the 1H NMR spectrum of the human inflammatory protein C5a. Biochemistry 27:3568.
15. Zuiderweg ERP, Nettesheim DG, Mollison KW, Carter GW (1989). Tertiary structure of human complement component C5a in solution from nuclear magnetic resonance data. Biochemistry 28:172.
16. Davis DG (1987). A novel method for determining internuclear distances and correlation times from NMR cross-relaxation rates. JACS 109:3471.
17. Zuiderweg ERP, Henkin J, Mollison KW, Carter GW, Greer J (1988). Comparison of model and nuclear magnetic resonance structures of the human inflammatory protein C5a. Proteins Str Funct Gen 3:139.
18. Huber R, Scholze H, Paques EP, Deisenhofer J (1980). Crystal Structure Analysis and molecular model of human C3a anaphylatoxin. Hoppe Seylers Z Physiol Chem 361:1389.
19. Nettesheim DG, Edalji RP, Mollison KW, Greer J, Zuiderweg ERP (1988). Secondary structure of complement component C3a anaphylatoxin in solution as determined by NMR spectroscopy: differences between crystal and solution conformations. Proc Nat Acad Sci USA 85:5036.

20. Chenoweth DE, Hugli TE (1978). Demonstration of specific C5a receptor on intact human polymorpho-nuclear leukocytes. J Immunol 120:109.
21. Edalji RP, Mollison KW, Zuiderweg ERP, Fey TA, Krause RA, Conway RG, Miller L, Lane B, Henkin J, Greer J, Carter GW (1987). Truncation of the amino or carboxyl terminus of recombinant C5a reduces receptor binding and biologic activity. Fed Proc 46:980.
22. Mollison KW, Mandecki W, Zuiderweg ERP, Fayer L, Fey TA, Krause RA, Conway RG, Miller L., Edalji RP, Shallcross MA, Lane B, Fox JL, Greer J, Carter GW (1989). Identification of receptor binding residues in the inflammatory complement protein C5a by site-directed mutagenesis. Proc Natl Acad Sci USA 86:292.
23. Reprinted with permission from Zuiderweg ERP, Nettesheim DG, Mollison KW, Carter GW (1989) Biochemistry 28:172 Copyright (1989) American Chemical Society.

Frontiers of NMR in Molecular Biology, pages 89-98

APPLICATIONS OF HETERONUCLEAR THREE DIMENSIONAL NMR SPECTROSCOPY

Stephen W. Fesik, Erik R.P. Zuiderweg,
Robert T. Gampe, Jr., and Edward T. Olejniczak

Pharmaceutical Discovery Division, Abbott Laboratories
Abbott Park, Illinois 60064

ABSTRACT Heteronuclear three-dimensional NMR spectroscopy has been applied in the study of a uniformly ^{15}N-labeled enzyme, CMP-KDO synthetase (MW=27,500), complexed with an inhibitor and CTP. Using this 3D technique, the NOESY spectrum of the ternary complex, which was extremely complex due to severe spectral overlap, was markedly simplified by the 3D technique by editing with respect to the ^{15}N chemical shifts. We have also found that heteronuclear 3D NMR experiments can be applied to small molecules with ^{13}C at natural abundance. This has been demonstrated in a 3D HMQC-COSY experiment on a concentrated solution of the aminoglycoside, kanamycin A. Ambiguities in the assignments which arise from protons resonating at the same frequency are resolved in this experiment by editing the COSY data by the ^{13}C frequencies.

INTRODUCTION

We and others have recently proposed (1,2) the use of heteronuclear three-dimensional NMR spectroscopy for resolving overlap observed in two-dimensional NMR spectra. In these experiments, a heteronuclear shift correlation and a homonuclear two-dimensional NMR experiment (e.g. COSY, NOESY) are combined. As previously demonstrated using a uniformly ^{15}N-labeled peptide (1), homonuclear COSY and NOESY spectra can be simplified using this technique by editing with respect to the heteronuclear chemical shifts. This method has the advantage over homonuclear 3D NMR in that a large heteronuclear J-coupling is involved in one of the coherence transfer steps resulting in an experiment with high sensitivity even when applied to large biomolecules.

In this report, we demonstrate the utility of heteronuclear three-dimensional NMR spectroscopy for simplifying the 2D NOE spectrum of a uniformly ^{15}N-labeled enzyme, CMP-KDO synthetase (248 amino acids MW=27,500) (3), complexed with an inhibitor and CTP. In addition, we show that heteronuclear 3D NMR can also be applied to small molecules with ^{13}C at natural abundance. This is illustrated in a 3D HMQC-COSY experiment on a concentrated solution of the aminoglycoside, kanamycin A (4).

METHODS

The ^{15}N-labeled tripeptide was prepared as previously described (1). For the HMQC-NOESY experiment, a 15 mM solution of the peptide in tetramethylene-d_8 sulphone was employed. Uniformly ^{15}N-labeled CMP-KDO synthetase was isolated as previously described for the unlabeled enzyme (3) from an overproducing strain of E. coli grown on $^{15}NH_4Cl$ and a nutrient broth of ^{15}N-labeled amino acids. The NMR sample contained an equimolar mixture (1.5 mM) of [^{15}N]CMP-KDO synthetase, inhibitor, and CTP in H_2O/D_2O (9/1) (pH=7.4). The kanamycin A NMR sample was prepared by dissolving the aminoglycoside (78 mM) in D_2O (pD=2.4).

NMR experiments were acquired on a General Electric GN500 or Bruker AM500 NMR spectrometer. Some of the acquisition parameters are given in the figure legends. All of the NMR data were processed in the format of the FTNMR program (Hare Research; Woodinville, WA) with a CSPI minimap array processor interfaced to a Vax 8350 computer. Contour maps of the 3D data sets were displayed on a Silicon Graphics workstation.

RESULTS

Figure 1 depicts the pulse sequences of two heteronuclear 3D NMR experiments (1). Both sequences begin with a heteronuclear multiple-quantum correlation (HMQC) experiment

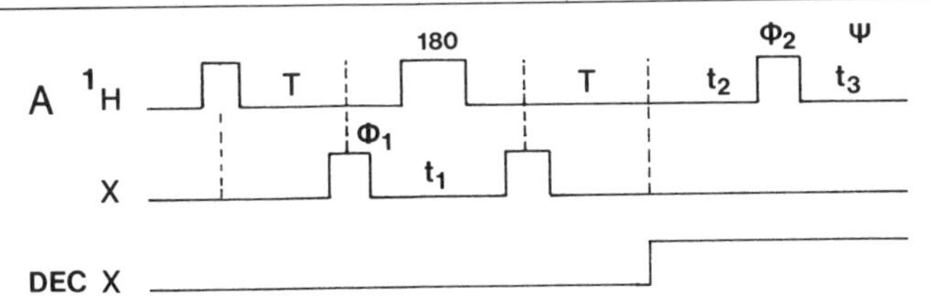

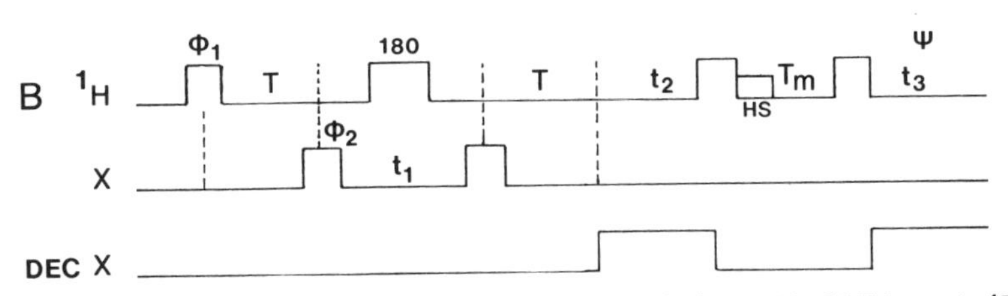

FIGURE 1. Pulse sequences of the (A) HMQC-COSY and (B) HMQC-NOESY experiments. [Source: S.W. Fesik and E.R.P. Zuiderweg (1)].

in which the heteronuclear frequencies evolve in t_1, and the frequencies of the protons scalar coupled to the heteronuclei evolve in t_2. Depending on the experiment, these proton signals in t_2 are converted to COSY (Fig. 1A) or NOESY (Fig. 1B) type responses. Thus, a heteronucleus that is frequency labeled in t_1 can be correlated to the frequency of its attached proton (t_2) and further correlated to a third proton (t_3) due to a scalar (COSY, TOCSY, relay) or dipolar (NOESY,ROESY) coupling (1).

Figure 2 depicts an HMQC-NOESY spectrum of an ^{15}N-labeled tripeptide. In this experiment, the transfer of magnetization is: $^{15}N(t_1) \rightarrow {}^{15}NH(t_2) \rightarrow {}^{15}NH(t_3) + C\alpha H(t_3)$ + side-chain protons (t_3). The ^{15}N-filtered NOESY spectrum of the tripeptide is shown on top. As shown in Figure 2, this spectrum is edited in ω_1 with respect to the ^{15}N frequencies of the peptide. Vertical lines, parallel to ω_1, connect the diagonal peaks of the amide protons in the top spectrum with the corresponding peaks located in the 3D data set at the

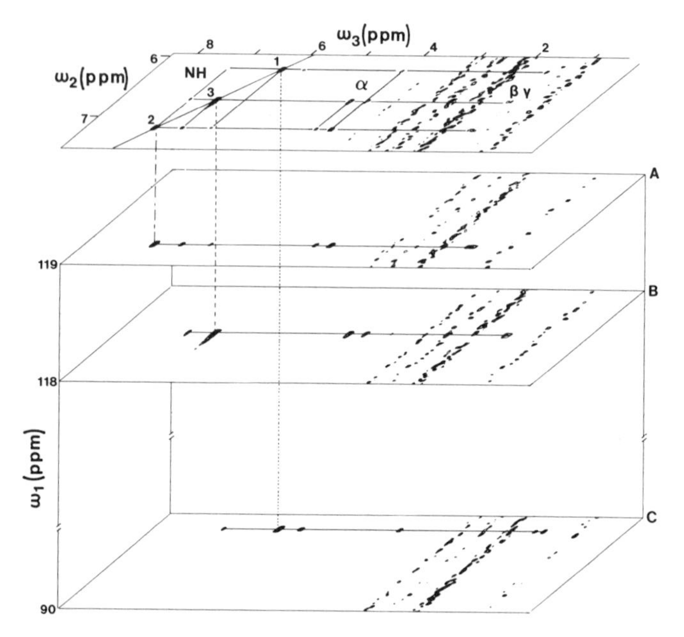

FIGURE 2. 3D HMQC-NOESY spectrum of a ^{15}N-labeled tripeptide collected as a series of 38 complex homonuclear NOESY experiments in which 32 complex t_2 values of 512 complex points (t_3) were obtained. Spectral widths of +/-1250, 400, and 3333 Hz were used in ω_1, ω_2, and ω_3, respectively. [Source: S.W. Fesik and E.R.P. Zuiderweg (1)].

^{15}N frequencies (ω_1) in the individual planes. The horizontal lines in the figure show the NOEs involving the amide protons that are scalar coupled to ^{15}N signals which resonate in different planes along ω_1 (1).

Heteronuclear 3D NMR experiments can also be performed in the reverse order, i.e. first the homonuclear 2D NMR experiment followed by the HMQC experiment (2,5,6). Figure 3 depicts an example of a pulse sequence for such an experiment. In this case, a NOESY and HMQC pulse sequence are combined to yield a 3D NOESY-HMQC experiment. Unlike in the HMQC-NOESY experiment (1), the full proton spectral width is indirectly detected in t_1,

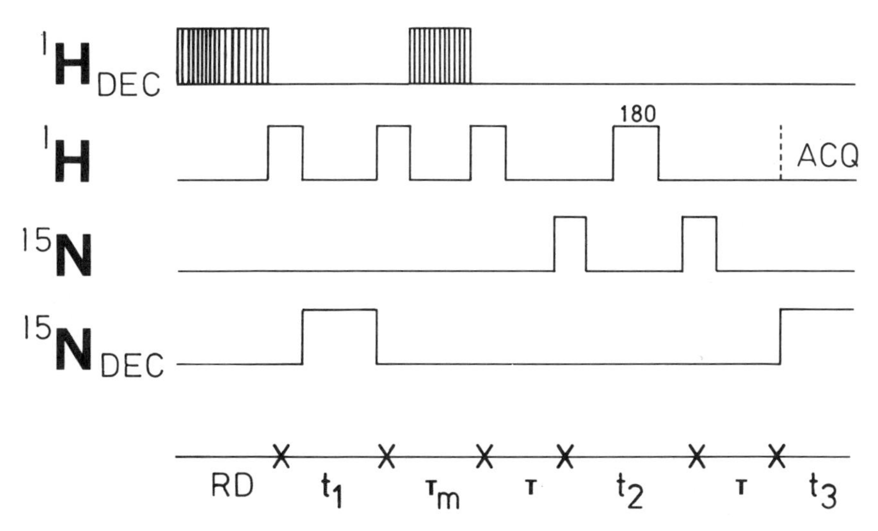

FIGURE 3. Pulse sequence of the 3D NOESY-HMQC experiment. [Source: E.R.P. Zuiderweg and S.W. Fesik (5)].

thereby requiring many t_1 points to achieve a high digital resolution. However, since the NOEs are indirectly detected in the NOESY-HMQC experiment, NOE cross peaks close to the large water signal can more easily be resolved.

For complicated NOESY spectra, the spectral simplification that can be achieved in this 3D experiment should dramatically improve the resolution of NOE cross peaks. This was demonstrated in a 3D NOESY-HMQC experiment performed on the inflammatory protein, C_{5a} (MW 8500), uniformly labeled with ^{15}N (5). The improved resolution allowed additional internuclear distances to be obtained from otherwise overlapping NOE cross peaks. In addition, ^{15}N resonance assignments were easily obtained (5) from the 3D data on C_{5a} which will be useful for studying the dynamics of this protein.

An important characteristic of heteronuclear 3D NMR is that a large one bond J-coupling is involved in one of the coherence transfer steps, suggesting that this technique will be applicable for NMR studies of proteins even larger than C_{5a}. Indeed, we have applied the 3D NOESY-HMQC experiment to a uniformly ^{15}N-labeled enzyme, CMP-KDO synthetase (MW=27,500), complexed with an inhibitor an CTP (6). As shown in the conventional 2D NOE spectrum of the

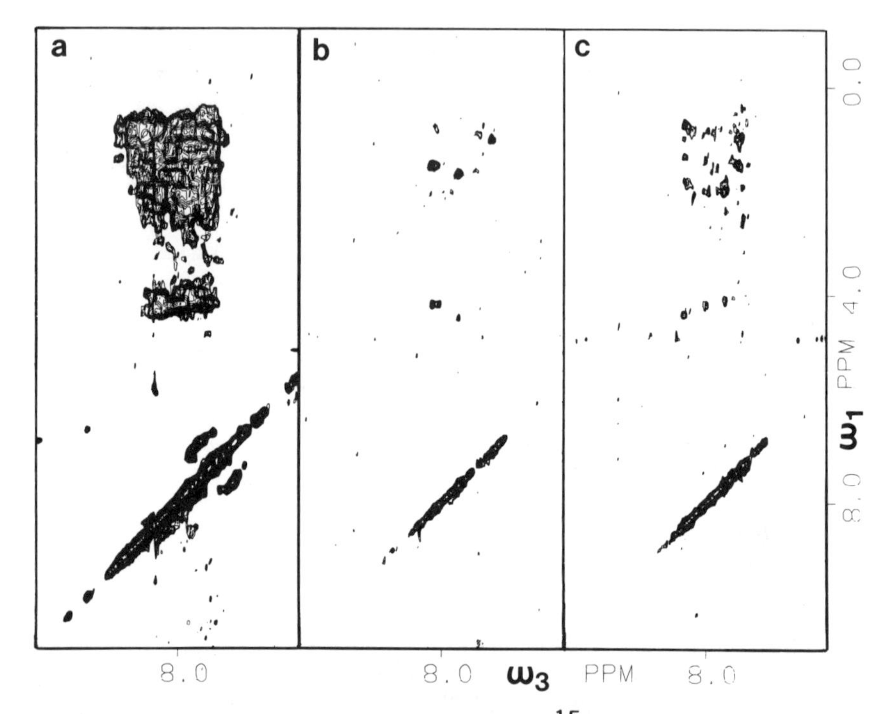

FIGURE 4. Contour plot of (a) an ^{15}N isotope-edited 2D NOE spectrum and (b,c) individual ω_1,ω_3 planes selected from a 3D NOESY/HMQC data set of a CMP-KDO synthetase/inhibitor/CTP complex [Source: S.W. Fesik et. al. (6)].

ternary complex (Figure 4a), the data is intractable since the individual NOE cross peaks cannot be resolved. However, in the ω_1,ω_3 planes of the 3D NOESY-HMQC experiment (e.g. Figures 4b,c), the NOE spectra are markedly simplified by editing with respect to the ^{15}N chemical shifts (ω_2). Thus, this 3D NMR experiment should provide a means to assign the proton NMR signals and possibly determine the structures of larger molecules which could not be accomplished by conventional 2D NMR techniques.

It is also possible to effectively apply heteronuclear 3D NMR experiments to studies of small molecules with ^{13}C at natural abundance. This has been demonstrated (7) in a 3D HMQC-COSY experiment on a concentrated solution of the aminoglycoside, kanamycin A. Although the COSY spectrum (Figure 5) of this aminoglycoside is quite complex in the

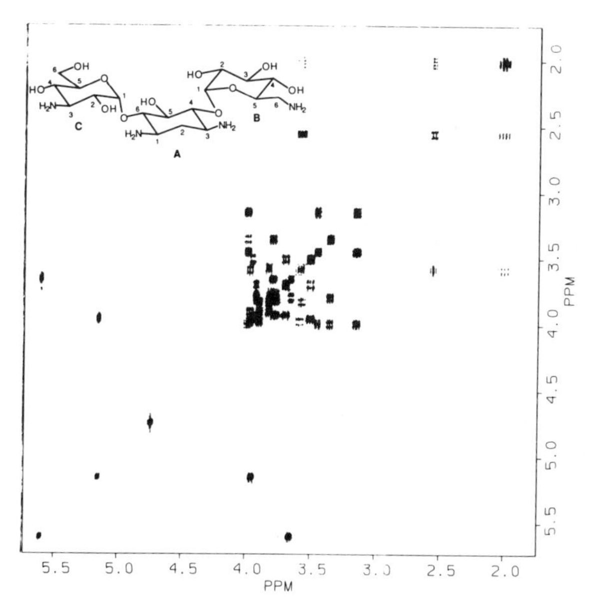

FIGURE 5. COSY spectrum of kanamycin A (structure shown in inset) [Source: S.W. Fesik et. al. (7)].

spectral region between 3.0-4.0 ppm, the individual ω_2,ω_3 planes (Figure 6) extracted from the 3D HMQC-COSY data set are markedly simplified. Only diagonal peaks corresponding to the protons attached to the carbons resonating at the ^{13}C frequencies selected in ω_1 and cross peaks corresponding to their scalar coupled partners are observed in each plane, greatly facilitating the analysis of the data (7). A three-dimensional contour map of the HMQC-COSY experiment of kanamycin A is shown in Figure 7. The figure illustrates how the $^1H-^1H$ COSY (top projection) or HMQC data (side projection) can be spread out in three dimensions and thereby simplified. Ambiguities in the assignments which arise from protons resonating at the same frequency are resolved in this 3D experiment by editing the COSY data by the ^{13}C frequencies. Analogous to a two-dimensional heteronuclear relay experiment, the 3D HMQC-COSY experiment provides relay information by correlating the chemical shifts of a carbon, the attached proton(s), and their scalar coupled partners. In contrast to the heteronuclear relay experiment, however, the 3D experiment uniquely defines the frequency of the relay spin which is important for resolving spectral overlap when ^{13}C signals resonate at identical frequencies (7).

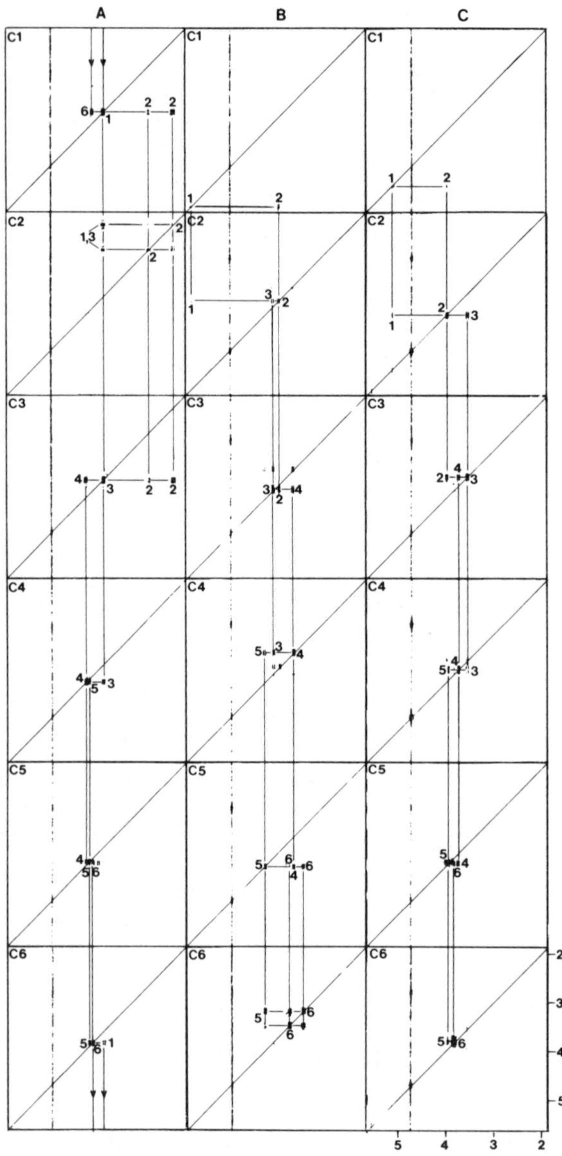

FIGURE 6. Individual planes (ω_2 vertical axis; ω_3, horizontal axis) of a 3D HMQC-COSY experiment at the ^{13}C frequencies (ω_1) corresponding to each of the carbon signals of 78 mM kanamycin A at ^{13}C natural abundance. The horizontal lines connect the protons that are scalar coupled, and the vertical lines connect the related planes. [Source: S.W. Fesik et. al. (7)].

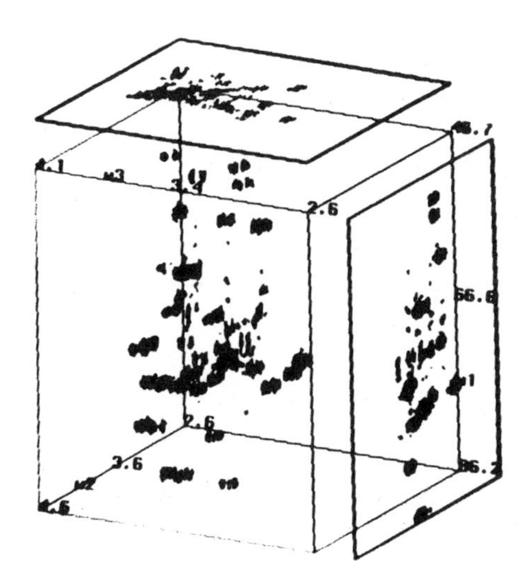

FIGURE 7. Three-dimensional contour map of a HMQC-COSY experiment of kanamycin A. The ω_2, ω_3 COSY (top) and ω_1,ω_2 HMQC projections (side) are displayed.

DISCUSSION

As demonstrated here, the two-dimensional NMR spectra of large isotopically labeled molecules can be dramatically simplified by editing with respect to the ^{15}N chemical shifts in a heteronuclear 3D NMR experiment. The high sensitivity achieved in these experiments is due to an efficient coherence transfer involving a large one-bond heteronuclear J coupling. Indeed, heteronuclear 3D NMR experiments are practical experiments that can be performed at low concentrations, providing that the molecule can be isotopically labeled. Due to the increase in resolution offered by this method, heteronuclear three-dimensional NMR spectroscopy will enable ^{1}H and ^{15}N NMR assignments of molecules that are too large to be assigned by conventional 2D NMR techniques. Furthermore, additional proton-proton distance constraints could be obtained by this technique from NOE cross-peaks that are not resolved in 2D NOE spectra, improving the ability of NMR to determine three-dimensional structures.

In addition, heteronuclear 3D NMR can be used to study small molecules at ^{13}C natural abundance. As illustrated for kanamycin A, the 3D HMQC-COSY spectrum was useful for resolving spectral overlap in all frequency domains which facilitated the analysis of the spectra. The increase in resolution afforded by this technique will help to automate peak-picking and assignment procedures and facilitate the extraction of J couplings (HMQC-COSY) and quantitative NOE information (HMQC-NOESY) used in the determination of three-dimensional structures (7).

ACKNOWLEDGMENTS

We thank Jim McAlpine for the sample of kanamycin A and Debra Weigl and Bill Kohlbrenner for help in the purification of CMP-KDO synthetase.

REFERENCES

1. Fesik SW, Zuiderweg ERP (1988). Heteronuclear three-dimensional NMR spectroscopy. A strategy for the simplification of homonuclear two-dimensional NMR spectra. J Magn Reson 78:588.
2. Marion D, Kay LE, Sparks SW, Torchia DA, Bax A (1989). Three-dimensional heteronuclear NMR of ^{15}N labeled proteins. J Am Chem Soc (in press).
3. Goldman RC, Bolling TJ, Kohlbrenner WE, Kim Y, Fox JL (1986). Primary structure of CTP:CMP-3-deoxy-D-manno-octulosonate cytidylyltransferase (CMP-KDO synthetase) from Escherichia coli. J Biol Chem 261:15831.
4. Koyama G, Iitaka Y, Maeda K, Umezawa H (1968). The crystal structure of kanamycin. Tetrahedron Lett 1875.
5. Zuiderweg ERP, Fesik SW (1989). Heteronuclear three dimensional NMR spectroscopy of the inflammatory protein C_{5a}. Biochemistry (in press).
6. Fesik SW, Gampe RT, Zuiderweg ERP, Kohlbrenner WE, Weigl D (1989). Heteronuclear three-dimensional NMR spectroscopy applied to CMP-KDO synthetase (27.5 kD). Biochem Biophys Res Commun (in press).
7. Fesik SW, Gampe RT, Zuiderweg ERP (1989). Heteronuclear three-dimensional NMR spectroscopy. Natural abundance ^{13}C chemical shift editing of ^{1}H-^{1}H COSY spectra. J Am Chem Soc 111:770.

Frontiers of NMR in Molecular
Biology, pages 99-108

Calcium Regulation of Muscle Contraction; NMR Studies of the Interaction of Troponin C with Calcium and Troponin I

A. Patricia Campbell, Brian J. Marsden, Jennifer Van Eyk,
Robert S. Hodges, and Brian D. Sykes

Department of Biochemistry, and MRC Group of Protein Structure and Function, University of Alberta, Edmonton, Alberta, Canada, T6G 2H7

ABSTRACT Short range distance measurements from 2D NOESY experiments are the foundation of most modern NMR structure determinations. Obtaining longer range distances requires the use of paramagnetic agents. Since the lanthanides have been widely used as calcium analogs, these paramagnetic metals are ideally suited for distance measurements in calcium-binding proteins. Relaxation measurements using gadolinium have been used to determine the structure of a peptide representing calcium-binding site III of troponin C, and to position a peptide representing the inhibitory region of troponin I on troponin C.

INTRODUCTION

Calcium acts as a cytosolic second messenger in many biochemical systems. The regulation of vertebrate skeletal muscle contraction involves the binding of calcium to a protein complex, troponin-tropomyosin, on the thin filament. Troponin consists of three subunits, troponin C (TnC), troponin I (TnI) and troponin T. The initial event in the contraction is considered to be the binding of Ca(II) to TnC, which then undergoes a conformational change which is transmitted through the troponin-tropomyosin complex to the thin muscle filament, culminating in the sliding of the thin and thick filaments past each other. The Ca(II)-mediated interaction between TnI and TnC leads to neutralization of the inhibitory activity of TnI on actomyosin ATPase. TnC has two classes of Ca(II)-binding sites; two high affinity sites, and two lower affinity Ca(II) selective sites. Each Ca(II) binding site exists in a helix-loop-helix motif[1], with 6 coordinating residues of the metal ion found within a central 12-residue loop flanked on both sides by α-helical regions.

To simplify the approach to studying the interaction of TnC with both Ca(II) and TnI, we have chemically synthesized peptides spanning sequences of interest on both these proteins. The first of these peptides is a 13-residue peptide spanning residues 103-115 (site III) on rabbit skeletal TnC, representing just the calcium-binding loop region with its six metal coordinating ligands at positions 1,3,5,7,9 and 12 in the loop. The second of these peptides is a synthetic analogue of the inhibitory region of rabbit skeletal TnI spanning residues 104-115, which represents the minimum sequence necessary for inhibition of actomyosin ATPase. This portion of TnI binds to TnC and is responsible, in part, for the Ca(II)-sensitive control exhibited by troponin[2].

^{1}H-NMR methods have been used to determine the structure of the TnC(103-115) metal ion complex, to determine the structure of the TnI(104-115) peptide when bound to TnC, and to locate the position on TnC where TnI(104-115) binds. While these studies have made extensive use of two-dimensional NOESY and transferred NOESY techniques, we have also used the paramagnetic lanthanide Gd(III) as a distance probe and these measurements will be the focus of this manuscript. Since only short range internuclear distances of 2-5 Å can be obtained via the nuclear Overhauser effect, the determination of longer range internuclear distances (7-15 Å) requires the use of paramagnetic agents such as the lanthanides. Since the lanthanides have been widely used as Ca(II) analogs, this approach is ideally suited to the study of calcium-binding proteins. In the first of the systems studied, Gd(III) was bound to TnC(103-115) and distances between residues in the loop and the central metal ion were determined from relaxation measurements, providing additional distances for the determination of the structure of the calcium-binding loop peptide metal ion complex. A variation of this scheme is offered in the second system studied where Gd(III) was bound to sites III and IV of intact TnC and distances determined to nuclei on the bound TnI peptide, thus providing a probe for the position of the interaction of TnI with TnC.

THEORY

The relaxation times T_1 and T_2 of a nucleus *i* near a paramagnetic site are represented by the Solomon-Bloembergen equations[3]:

$$1/T_{1Mi} = (A/r_i^{\,6})[3\tau_c/(1+(\omega\tau_c)^2)] \qquad \text{1a}$$

$$1/T_{2Mi} = (B/r_i^{\,6})[4\tau_c + 3\tau_c/(1+(\omega\tau_c)^2)] \qquad \text{1b}$$

where τc is the correlation time, ω is the Larmor precession frequency and r_i is the distance between the nucleus i and the paramagnetic ion.

If we consider a paramagnetic system in chemical exchange, the observed relaxation rates are $1/T_{1OBS}$, which have diamagnetic as well as paramagnetic contributions. If Gd(III) (4 f^7) is used in a paramagnetic system (7 unpaired electrons), La(III) (4 f^0) may be used as a suitable diamagnetic control (no unpaired electrons). Thus, for nucleus i, the paramagnetic contributions to the relaxation rates, $1/T_{1Pi}$ and $1/T_{2Pi}$ are given by:

$$[1/T_{1OBS}^{Gd} - 1/T_{1OBS}^{La}]^i = f_B/(T_{1Mi} + \tau_M) = 1/T_{1Pi} \qquad 2a$$

$$[1/T_{2OBS}^{Gd} - 1/T_{2OBS}^{La}]^i = f_B/(T_{2Mi} + \tau_M) = 1/T_{2Pi} \qquad 2b$$

where f_B is the 'fraction bound' and τ_M is the exchange lifetime of the bound state.

To calculate distances from $1/T_{1Pi}$ and $1/T_{2Pi}$ measurements requires both knowledge of τ_M to separate out $1/T_{1Mi}$ and $1/T_{2Mi}$, and knowledge of τ_C. Determination of τ_C and τ_M has been previously outlined by Marsden etal.[4] using the following expression:

$$T_{2Pi} = T_{1Pi} (1/\acute{A}) + \tau_M (\acute{A}-1)/\acute{A} \qquad 3a$$

$$\text{where} \quad \acute{A} = 7/6 + (2/3)(\omega\tau_C)^2 \qquad 3b$$

Thus, a plot of T_{2Pi} vs T_{1Pi} for nuclei i = 1,...,n yields a slope of $1/\acute{A}$ and an intercept of $\tau_M (\acute{A} -1)/\acute{A}$. The slope yields τ_C from equation 3b; and τ_M may be obtained from the intercept. Then r_i may be calculated from equation1a.

STRUCTURE OF TROPONIN-C CALCIUM-BINDING LOOP PEPTIDES

We have previously studied the interaction of metal ions with a wide variety of synthetic peptide analogs of the calcium-binding loop regions of TnC, in an attempt to understand the variation in Ca(II)/Mg(II) specificity and affinity of these metal binding sites[5]. For example, we have synthesized analogs which represent all possible permutations and combinations of Asp and Asn in the first three liganding positions in these loops, and concluded on the basis of these studies that the metal affinity is a balance of attractive forces between the ligand and the metal and repulsive forces between the negatively charged ligands so that the relative spatial orientation of the charges is important. It is important to demonstrate, however, that the peptide-metal ion complex adopts the same structure in solution as the same sequence has or would have in the protein for these conclusions to be valid.

In this paper we have chosen the peptide shown below, which is an analog of site III of rabbit skeletal TnC (positions 103 to 115 in the sequence) with Asp in place of Asn in position 105, since it has high metal ion affinity:

Ac-ASP-ARG-ASP-ALA-ASP-GLY-TYR-ILE-ASP-ALA-GLU-GLU-LEU-NH_2

103	105	107	109	111	114
X	Y	Z	-Y	-X	-Z

We have approached the determination of the structure of the metal-ion complex of this peptide (D105) using 2D NOESY internuclear distance measurements, and distance measurements using the paramagnetic metal Gd(III). Gd(III) is a calcium analog which binds in the place of Ca(II) but with higher affinity because of the increased charge.

The ^{1}H NMR spectrum of the diamagnetic La(III) complex of this peptide has been completely assigned elsewhere[4]. When very small amounts of Gd(III) are added to the peptide-La(III) complex there is a very large increase in the observed relaxation rates $1/T_1$ and $1/T_2$. Since the observed linewidths are related to $1/T_2$ ($\Delta\nu = 1/T_2$), this is most obvious as linebroadening and shown for the methyl resonances in Figure 1. The increase in spin-lattice relaxation rate as a function of added Gd(III) for selected resonances is shown in Figure 2.

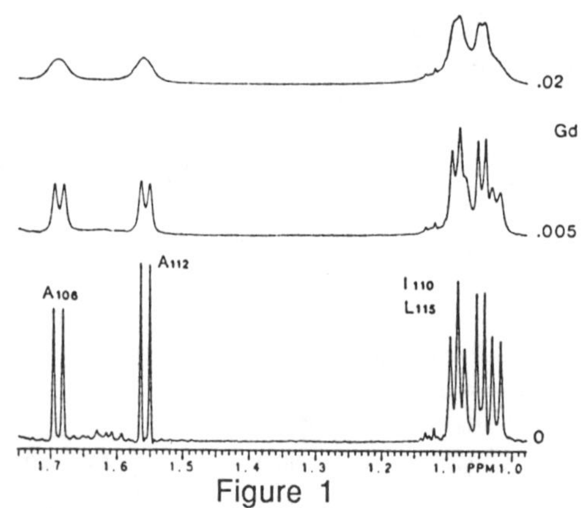

Figure 1

Methyl region of ^{1}H NMR spectrum of La(III)-complex of D105 TnC peptide as a function of added Gd(III) as a fraction of total Ln(III).

The slopes of the lines presented in Figure 2 give the values of the paramagnetic contribution to the spin-lattice relaxation in the Gd(III)-peptide complex ($1/T_{1Pi}$). When similar plots are made from the linewidth measurements, the values of T_{1Pi} and T_{2Pi} can be plotted versus one another as shown in Figure 3. This plot yields a value for the exchange lifetime of the lanthanide-D105 complex ,τ_M , of 2 x 10^{-4} s; and a value of the correlation time for the interaction, τ_C , of 0.7 x 10^{-9} s using equation 3. The distances of the assigned nuclei to the metal can then be calculated from the spin-lattice relaxation rates and these values using equation 1.

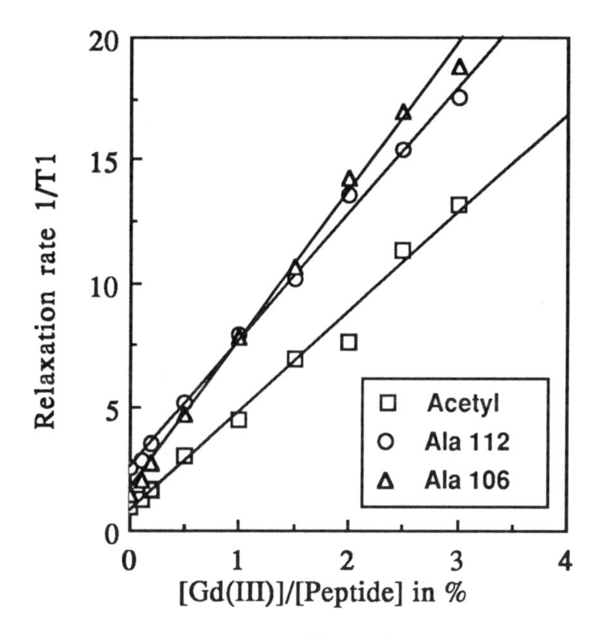

Figure 2
Spin-lattice relaxation rate of selected resonances in La(III)-complex of D105 TnC peptide as a function of added Gd(II).

The distances obtained from NMR measurements (including both the 2D NOESY measurements and the Gd relaxation measurements) are plotted against the nearest equivalent distances

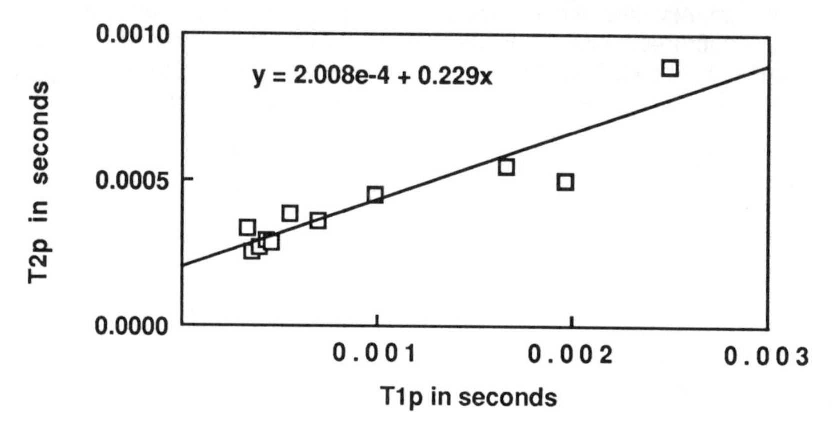

Figure 3
Plot of paramagnetic contribution to T_2 versus T_1 for protons of D105-Gd(III) complex (Adapted from reference 4)

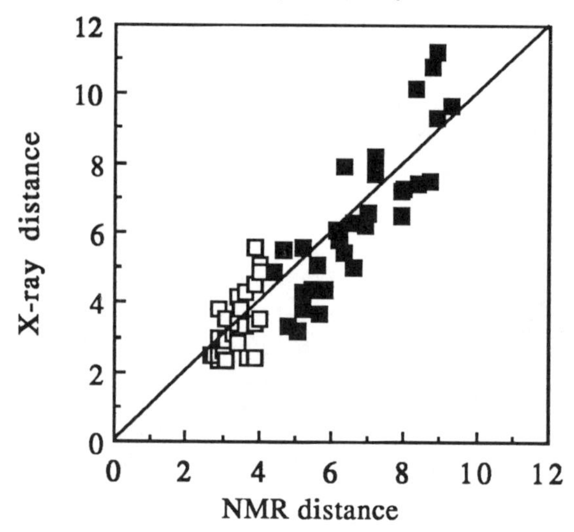

Figure 4
Plot of distances from NMR (nOe,open squares;Gd, filled) versus X-ray (adapted from reference 4)

(the sequences are not identical) obtained from the X-ray structure of turkey skeletal TnC[6] in Figure 4. This plot shows clearly the different ranges of distances obtainable from the two types of NMR measurements, and the essential similarity between the solution and X-ray structures. The RMS different between the NMR and X-ray distances is 1.1 Å. This is considered excellent in the present example since there is no *a priori* reason that, even if the structures are identical in all important aspects, all distances will be the same; for example, the side chain of Ile is buried in the interior of the protein but exposed to solvent in the peptide.

THE INTERACTION OF TnC WITH TnI INHIBITORY PEPTIDE

The structure of the TnI peptide bound to TnC has been determined using a technique called the transferred nuclear Overhauser effect (trnOe)[7,8]. The trnOe allows the transfer via chemical exchange of conformational information between the bound and free peptide. In the unbound form, the peptide tumbles rapidly in solution and the nOes are small and positive; whereas, when bound to TnC, the peptide tumbles slowly with the protein and the peptide nOes are large and negative. Thus, in the presence of TnC,negative nOes are measured on the easily detected free ligand resonances, giving conformational information of the bound state. The bound structure of the peptide has been calculated using the computational techniques of restrained energy minimization and molecular dynamics[9] wherein the experimentally determined distance restraints, r_{ij}s, are used as pseudo potential energy terms.

The theory of the trnOe and the structure of the bound peptide have been published elsewhere[10]. The bound TnI peptide contains two "bends" in the structure around residues Gly104-Arg108 and Arg112-Arg115, which bring the hydrophobic side chains of Phe106, Leu111 and Val114 into close proximity. This arrangement defines a hydrophobic core and presents a hydrophobic face, spanning Pro109-Pro110, capable of interacting with a hydrophobic patch on the surface of TnC. The peptide contains 4 arginines (Arg108, Arg112, Arg113 and Arg115) and 2 lysines (Lys105 and Lys107), and therefore is highly basic. These bulky and charged side chains extend out from the core of the structure and may either interact with the solvent or with acidic residues on TnC.

The position of the TnI peptide on TnC has been studied using NMR relaxation studies with Gd(III) in sites III and IV of whole TnC. Based on studies of biologically active fragments of TnI and TnC [2,11], the sequence of TnI spanned by residues 104-115 is presumed to bind near the N-terminal helix of site III on TnC. By measuring increases in $1/T_{1M}$ for assigned nuclei on the TnI peptide (a result of the proximity of the peptide to Gd(III) in site III and/or site IV of TnC), one can determine the distances, r_{iM}s between these specific nuclei and sites III and IV on TnC. Figure 5 shows the increase in the relaxation rate, $1/T_{1OBS}$, for the TnI peptide Leu111 δCH3 methyl group as a function of the fraction of peptide bound to lanthanide (La(III) or Gd(III)) saturated TnC.

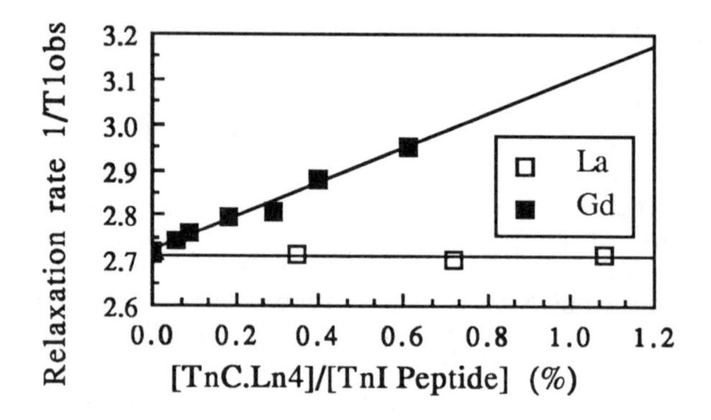

Figure 5
Spin lattice relaxation rates of Leu 111 δCH3 protons as a function of TnC-La(III)$_4$ or TnC-Gd(III)$_4$ concentration

Figure 6 shows the plot of T_{2Pi} versus T_{1Pi} for nuclei on the TnI peptide. An inspection of the graph reveals that although the intercept { τ_M (Á -1)/Á } is readily obtained as $(4.0 \pm 2.0) \times 10^{-4}$ s, the slope {1/Á} is very shallow and cannot be measured with any accuracy. This is due to the relationship between Á and τ_C described in equation 3b; as the size of the complex increases, so does the correlation time τ_C, and likewise Á as the square of τ_C. Thus, for the calcium-binding peptide TnC(103-115), τ_C is small, Á is small, and the slope 1/Á is large and measurable. However, for the TnI(104-115)peptide:TnC complex, τ_C is large, Á is large, and the slope1/Á is small and difficult to measure. In this case, limits mustbe placed on the value of τ_C and Á . It is clear from Figure 6 that Á>100, which from equation 3b limits $\tau_C > 4 \times 10^{-9}$ s. The maximum value of τ_C is equal to the rotational correlation time, τ_R, which can be readily calculated from the Stokes-Einstein equation. Approximating TnC as a spherical particle gives $\tau_R = 20\times10^{-9}$ s. Thus with the limits of defined as 4×10^{-9} s < τ_C < 20×10^{-9} s, and $\tau_M = 4\times10^{-4}$ s (for small $\acute{A}$, intercept= $\tau_M(\acute{A}-1)/\acute{A} = \tau_M$), a range of distances may be calculated. Table 1 shows the range of distances, calculated for various protons on the TnI peptide.

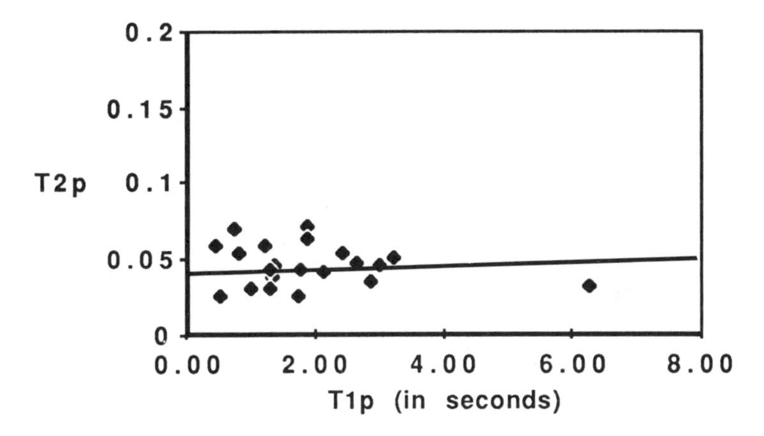

Figure 6
Plot of paramagnetic contributions to relaxation times for TnI peptide bound to $Gd(III)_4$-TnC

Table 1
$TnC-Gd(III)_4$ to TnI(104-115) distances

RESIDUE	SPIN SYSTEM	R(Å)MIN	R(Å) MAX
104	Gly αCH2	7.00	9.18
105	Lys γCH2	8.16	10.69
106	Phe αCH	6.40	8.39
106	Phe βCH2	6.71	8.80
106	Phe Ring protons	5.98	7.84
107	Lys γCH2	8.06	10.56
108	Arg αCH2	7.46	9.78
109	Pro γCH2	9.29	12.17
110	Pro γCH2	7.80	10.22
111	Leu (δCH3)2	7.88	10.32
114	Val (γCH3)2	7.84	10.27

The final step in the positioning of the TnI peptide on TnC will be docking of the bound conformation of the TnI peptide to the solved crystal structure of turkey-TnC[6] using the r_{iM}'s generated from the Gd(III) relaxation measurements, in conjunction with energy miminization programs to optimize electrostatic, hydrophobic and Van der Waals contacts between the peptide and the protein.

ACKNOWLEDGEMENTS Supported by the Medical Research Council of Canada and the Alberta Heritage Foundation for Medical Research .

REFERENCES

1. Kretsinger, RH, Nockolds, CE (1973). Carp Muscle Calcium-Binding Protein. J Biol Chem 248:3313.
2. Van Eyk, JE, Hodges, RS (1988). The Biological Importance of Each Amino Acid Residue to the Troponin I Inhibitory Sequence 104-115 in the Interaction with Troponin C and Tropomyosin-Actin. J Biol Chem 263:1726.
3. Bloembergen, N (1957). Proton Relaxation Times in Paramagnetic Solutions. J Chem Phys 27:572.
4. Marsden, BJ, Hodges, RS , Sykes BD (1989).A 1H NMR Determination of the Solution Conformation of a Synthetic Peptide Analog of Calcium-Binding Site III of Rabbit Skeletal Troponin C. Biochem ,in press.
5. Marsden, BJ, Hodges, RS, Sykes, BD (1988). 1H NMR Studies of Synthetic Peptide Analogues of Calcium-Binding Site III of Rabbit Skeletal Troponin-C: Effect of the Lanthanum Affinity of the Interchange of Aspartic Acid and Asparagine Residues at the Metal Ion Coordinating Positions. Biochem 27:4198.
6. Herzberg, O. James, MNG (1988). Refined Crystal Structure of Troponin C from Turkey Skeletal Muscle at 2.0 Angstrom Resolution. J Mol Biol 203:761.
7. Clore, GM, Gronenborn, AM (1982) Theory and Applications of the Transferred Nuclear Overhauser Effect to the Study of the Conformations of Small Ligands Bound to Proteins. J Magn Reson 48:402.
8. Clore, GM, Gronenborn, AM (1983) Theory and the Time Dependent Transferred Nuclear Overhauser Effect: Applications to Structural Analysis of Ligand-Protein Complexes in Solution. J Magn Reson 53:423.
9. Van Gunsteren, WF, Berendsen HJC, Laboratory of Physical Chemistry, University of Groningen, Nijenborgh 16, 9747 Ag Groningen, The Netherlands.
10. Campbell, AP, Sykes BD (1989). Conformation of a Troponin-I Peptide Bound to Troponin-C. In Hidaka H (ed): "Sixth International Symposium on Calcium Binding Proteins in Health and Disease,". New York: Plenum Press, in press.
11. Grabarek, Z, Drabikowski, W, Leavis, PC, Rosenfeld, SS, Gergely J (1981). Proteolytic Fragments of Troponin C: Interactions with the Other Troponin Subunits and Biological Activity. J Biol Chem 256:13121.

Frontiers of NMR in Molecular
Biology, pages 109-118

NMR SPECTRA OF SYNTHETIC MEMBRANE BOUND COAT PROTEIN SPECIES[1]

K. Shon[2], P. Schrader[2], S. Opella[2], J. Richards[3], and J. Tomich[3,4]

[2]Department of Chemistry, University of Pennsylvania
Philadelphia, Pennsylvania 19104
[3]Division of Chemistry and Chemical Engineering,
California Institute of Technology, Pasadena, California 91125
[4]Division of Medical Genetics,
Childrens Hospital of Los Angeles,
4650 Sunset Blvd., Los Angeles, California 90027

ABSTRACT High resolution ^{1}H and ^{15}N NMR spectra of several synthetic membrane bound coat protein species in micelles demonstrate two major advantages of using solid-phase peptide synthesis to prepare protein samples for spectroscopic studies. First, solid-phase peptide synthesis allows preparation of large amounts of protein species that cannot be isolated and purified from bacterial growths. Secondly, solid-phase peptide synthesis allows the placement of isotopic labels in specific locations; this is in contrast to selective labelling with bacterial growths where all residues of the same type are labelled. NMR spectra of three different protein species obtained from solid-phase peptide synthesis are presented in this paper. The spectra are from a 23 residue fragment of M13 procoat protein that serves as a substrate for leader peptidase, the 46 residue Pf1 coat protein, and the 82 residue Pf1 procoat protein. These spectra demonstrate that sufficient quantities for NMR experiments can be obtained of moderately large protein species by solid-phase peptide synthesis. Specific labelling of the Tyr at position +25 of Pf1 coat and procoat proteins results in single line ^{15}N NMR spectra; this is in contrast to the selectively labelled species from bacterial growths with both Tyr +25 and Tyr +40 contributing resonances.

---

[1]This work was supported by National Institutes of Health grants AI-20770 and GM-16424.

INTRODUCTION

Two of the main advantages of studying protein samples prepared by solid-phase peptide synthesis have proven to be important in our initial studies of the insertion and processing of the procoat proteins from filamentous bacteriophages. These advantages are derived from the ability to prepare large amounts of protein species that are impossible to obtain from bacterial growths and the ability to place isotopic labels in specific locations, even when the same amino acid is present in many different locations in the sequence. There are several additional advantages to studying chemically synthesized proteins, including the ability to generate single and multiple site mutants without concern for proteolysis or purification and the ability to incorporate non-standard amino acids and even non-amino acids into specific locations (1).

The targeting of proteins to specific membranes and their insertion into or translocation across membranes constitutes one of the most important and topical areas of research pursued by cellular and molecular biologists. Perhaps the major mechanism for these processes involves leader sequences, 20-40 residue sections of the proteins often located at the N-terminus, that are removed proteolytically from the preprotein species to form the mature protein species (2).

In order to understand, and possibly control, protein compartmentalization, it is essential to describe the structure and dynamics of several protein species, including the preprotein and mature proteins. Biophysical studies of intact preproteins with their leader sequences attached have not been performed to date, largely because of the difficulty in isolating and purifying protein species that are present only transiently in the cell. In addition, membrane bound proteins are often difficult to study by biophysical methods capable of atomic resolution; in particular, few membrane bound proteins have been crystallized to date. NMR spectroscopy has the potential for describing these proteins, if they can be obtained in substantial quantities and labelled with stable isotopes.

Filamentous bacteriophages are nucleoprotein complexes (3). Even though no lipids are associated with the virus particles their coat proteins do interact with membranes. Newly assembled virus particles are extruded through the cell membrane with the coat protein going directly from the membrane environment, where it is stored prior to assembly, to the virus particles, where the protein surrounds the DNA. During the virus lifecycle, the coat protein is synthesized in the cell cytoplasm as the product of the viral gene 8 with a leader sequence attached to the N-terminus of the protein. After insertion of the intact procoat protein into the cell membrane, the leader sequence is removed by bacterial leader peptidase (4). This system is of considerable interest because the protein interacts with both membranes and DNA and because it can serve as a model for the insertion and processing of membrane proteins, since the bacterial protein species are small with 46-82 residues and are not glycosylated.

NMR spectroscopy has been utilized in studies of the coat proteins in membrane environments (5) and as the major structural element of the virus particles (6). Even though the proteins are small and can be labelled selectively and uniformly in biosynthesis, there remain resolution and assignment problems resulting from multiple residues of the same amino acid. These problems can be overcome by specific labelling of chemically synthesized coat proteins. More importantly, it has not been possible to isolate significant amounts of procoat proteins from bacterial growths (7). Intact procoat proteins from filamentous bacteriophages are within the range of current synthetic techniques at 73 (M13) and 82 (Pf1) residues. In addition, some protein domains or subsequences are of interest by themselves and can be synthesized.

Preliminary ^{1}H and ^{15}N NMR spectra of coat and procoat protein species are described in this paper. These spectra clearly demonstrate two of the main advantages of using chemically synthesized proteins in NMR studies, namely the preparation of large amounts of species that can be obtained from bacterial growths and the specific incorporation of stable isotopes into the various protein species. The spectra described in this paper were obtained on the membrane bound forms of the protein species in detergent micelles in solution. These same protein species are also being studied in phospholipid bilayers by solid-state NMR spectroscopy.

MATERIALS AND METHODS

Peptide Synthesis

All peptides were synthesized using PAM resins and N^{α} t-boc amino acids (Peninsula) using an automated solid-phase protocol, based on the principles outlined by Merrifield on an Applied Biosystems model 430 peptide synthesizer (1). This instrument has been modified to carry out 0.1 mmole syntheses using flow-through protocols. To insure a high degree of coupling efficiency, all amino acids were double coupled. The ^{15}N isotopic label, t-boc-^{15}N-tryrosine-BZ-ether (Cambridge

Isotope Laboratories), was coupled by hand using a two-fold excess of the amino acids following automated coupling of the first 20 residues of Pf1 coat protein. The reaction was allowed to proceed for 90 minutes and incorporation of the label (>95%) was measured by a quantitative ninhydrin assay (1). The resin was subsequently washed, dried and returned to the reaction vessel of the automated synthesizer. The resin was recoupled (no TFA) using unlabelled tyrosine to complete the coupling. The synthesis was then allowed to go to completion as described above. After addition of the final amino acid, the resin was dried *in vacuo* prior to HF cleavage. Cleavage and full deprotection of peptides was carried out using 0.25 g of the pretreated peptide-resin in the presence of 0.33 ml of scavenger (p-cresol), and 0.33 ml reducing agent (p-thiocresol), and 8-10 ml anhydrous HF (Matheson). Condensation of the HF was carried out at -40°C and the reaction was allowed to proceed for 1-1.5 hr at 0°C employing an Immuno-Dynamics HF cleavage apparatus. The HF was removed at the end of the reaction and the cleavage reaction mixture was extracted with 100 ml of cold anhydrous diethyl ether. The resultant precipitate, which contains the cleaved peptide and vinyl benzene resin, was collected and dried *in vacuo* over KOH.

The peptides were separated from the resin by dissolving the peptide precipitate in 5% acetic acid followed by filtering a second ether extraction. This material was concentrated, incubated with SDS and lyophilized. All peptides are analyzed by Edman degradation using an Applied Biosystems model 477 gas-phase protein sequenator as well as by RP-HPLC on a Vydac C_4 column using 0.1% TFA in a water/acetonitrile gradient containing 0.1% n-octyl glucopyranoside (8). The apparent over all yield of the peptides was >90% as determined by integration of the HPLC peak areas.

Sample Preparation

The samples for ^1H NMR experiments were prepared by dissolving the synthetic peptide containing residues -15 and +8 of M13 procoat protein in H_2O or D_2O in the presence of perdeuterated sodium dodecyl sulfate (SDS) (Merck Sharp & Dohme Isotopes). 7 mg of the peptide and 14 mg of SDS resulted in 6 mM solutions. The pH was adjusted to 4.0 and the samples lyophilized. The sample in H_2O was prepared by redissolving the lyophilized sample in 0.45 mL of 2 mM citrate buffer, and the same sample was again lyophilized and redissolved in "100%" D_2O for the comparison of the peptide in H_2O and D_2O.

For ^{15}N NMR experiments, 2 mM solutions of synthetic Pf1 coat and procoat proteins were prepared. 80 mg of recrystallized electrophoretic-grade SDS (Boehringer Mannheim) was added to 5% acetic acid aqueous solution containing 20 mg of the synthetic proteins. The pH of the sample was adjusted to 4.0 before lyophilization. The sample was then redissolved in 2 mL of 2 mM citrate buffer. The samples containing 10% D_2O were transferred to 10 mM NMR tubes. Biosynthetically ^{15}N labelled Pf1 coat protein was prepared the same as the synthetic proteins except for the absence of 5% acetic acid. ^{15}N was selectively introduced to two tyrosine sites in Pf1 coat protein by growing the phage in the medium with 99% ^{15}N labelled tysosine (Cambridge Isotopes) as the sole source of the labelled amino acid.

NMR Experiments

^{1}H spectra were obtained following single pulse excitation with weak presaturation irradiation at the H_2O resonance which was turned off during data acquisition. 2K data points were collected with spectral width of 5000 Hz. ^{5}N NMR spectra were obtained with the DEPT pulse sequence (9). 2K points were collected during data acquisition with a spectral width of 2000 Hz. ^{15}N chemical shifts are referenced externally to ^{15}N-acetylglycine amide resonance at 90.4 ppm. All spectra were obtained on a JEOL GX400WB spectrometer.

RESULTS

NMR spectra of three different protein species are presented in Figures 2 and 3. Relatively large amounts (>5 mg) of proteins are needed for NMR experiments. Coat protein itself can be readily obtained from infected bacteria in selectively or uniformly labelled forms in these amounts (5). However, procoat protein and the fragment encompassing residues -15 to +8 can only be obtained by chemical synthesis in sufficient amounts for NMR experiments.

Interesting comparisons can be between class I (M13, fd) and class II (Pf1) bacteriophages. Therefore, our initial results include spectra from species from both types of bacteriophages (3). Figure 1 contains the amino acid sequences of the procoat proteins from M13 and Pf1 bacteriophages. Residue +1 becomes the amino terminus of the mature coat protein (10). Although there are no apparent direct sequence homologies between these two proteins, they do follow very similar design principles. The mature coat proteins have a hydrophobic midsection flanked by acidic (C-terminal) and basic (N-terminal) hydrophilic sections and the leader sequences appear to have a central hydrophobic section as well. The hydrophobic sections are believed to form membrane spanning helices, with the N- and C-termini of the procoat proteins exposed to the cell cytoplasm, but only the C-termini of the mature coat proteins exposed to the cytoplasm since the N-termini are exposed to the periplasm. Initial results from three protein species derived from the sequences in Figure 1 are described in this paper. They are the intact Pf1 procoat protein (residues -36 to +46), Pf1 coat protein (residues +1 to +46), and a portion of M13 procoat protein (residues -15 to +8) that serves as a substrate for bacterial leader peptidase.

A. M13 Procoat Protein

MKKSLVLKASVAVATLVPMLSFA
-23 -15 -1

AEGDDPAKAAFDSLQASATEYIGYAWAMVVVIVGATIGIKLFKKFTSKAS
+1 +8 +50

B. Pf1 Procoat Protein

MKAMKQRIAKFSPVASFRNLSIAGSVTAATSLPAFAG
-36 -1

VIDTSAVESAITDGQFDMKAIGGYIVGALVILAVAGLIYSMLRKA
+1 +46

Figure 1. Sequences of (A) M13 (7) and (B) Pf1 (11) procoat proteins.

Figure 2 contains one-dimensional ^1H NMR spectra of the peptide containing residues -15 to +8 of M13 coat protein in SDS micelles in H_2O (Figure 2B) and in D_2O (Figure 2A). The downfield aromatic and amide N-H resonance regions are shown. The comparison of the amide N-H resonance intensities in H_2O and in D_2O solutions demonstrate that a reasonable fraction of the amide hydrogens exchange with solvent deuterons relatively slowly. The spectra in Figure 2 have two of the criteria necessary to demonstrate that a protein is in a stable folded conformation and these are the chemical shift dispersion and relatively slowly exchanging amide N-H hydrogens.

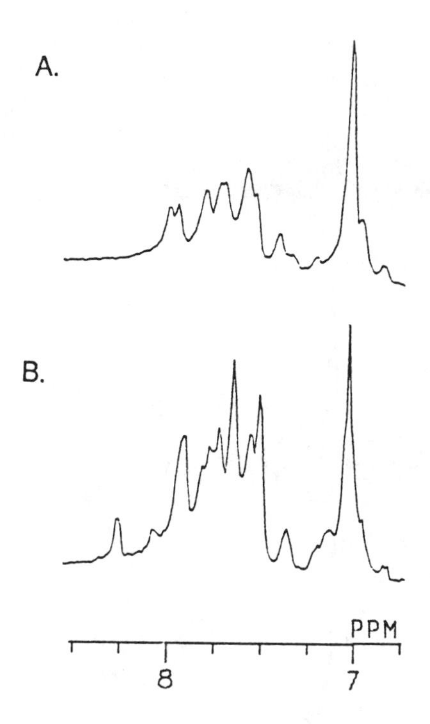

Figure 2. ^1H NMR spectra of peptide derived from M13 procoat protein (residues -15 to +8), in ^2H-SDS micelles at 35°C pH 4.0 in (A) D_2O and (B) H_2O.

The resonances in the spectra of Figure 3 arise from specifically or selectively ^{15}N labelled residues in the coat and procoat proteins of Pf1 bacteriophage. These spectra are on proteins samples in SDS micelles in H_2O solution and were obtained by polarization transfer of ^1H magnetization to the ^{15}N nuclei for sensitivity enhancement. As indicated in Figure 1, Pf1 procoat protein has tyrosine residues in positions +25 and +40. ^{15}N-Tyr can be readily incorporated into both sites of the protein biosynthetically. However, the coat protein species, but not the procoat protein species, can be isolated and purified from the bacterial growth. The spectrum in Figure 3C contains resolved single line resonances for each of the two tyrosines of

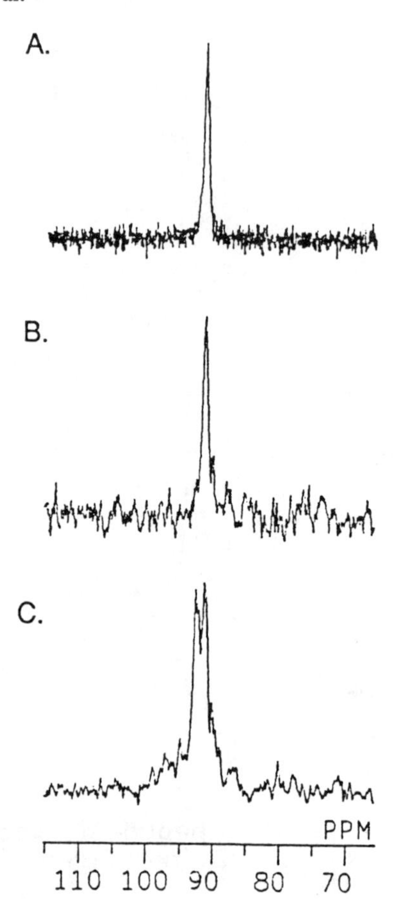

Figure 3. ^{15}N spectra of Pf1 coat proteins in SDS micelles in H_2O at 50°C, pH = 4.0. (A) Specifically ^{15}N labelled synthetic Pf1 (Tyr +25) coat protein. (B) Specifically ^{15}N labelled synthetic Pf1 (Tyr +25) procoat protein. (C) Selectively ^{15}N labelled viral Pf1 (Tyr +25 and Tyr +40) coat protein.

Pf1 coat protein; since the protein is selectively labelled with ^{15}N-Tyr there is no way to differentiate between the resonances from ^{15}N in residue +25 or in residue +40. The spectrum in Figure 3B is also of Pf1 coat protein, however this protein sample was prepared by solid-phase chemical synthesis with ^{15}N-Tyr incorporated into residue +25 and natural abundance ^{14}N-Tyr incorporated into residue +40. The comparison of the spectra in Figures 3B and 3C assigns the ^{15}N-Tyr resonances to specific residues in Pf1 coat protein. The spectrum in Figure 3C with a resonances from only a single residue also can be used for a variety of spectroscopic studies that do not have chemical shift resolution, including the analysis of motionally averaged powder patterns in solid-state NMR spectra of proteins in phospholipid bilayers (5).

The spectrum in Figure 3A demonstrates the power of chemical synthesis for NMR studies of proteins. The protein is Pf1 procoat protein (residues -36 to +46) that can not be isolated in significant quantities from infected cells. And it is specifically labelled at residue +40 with ^{15}N-Tyr. The single line resonance from Tyr +25 in the ^{15}N NMR spectrum of Pf1 procoat protein has very similar properties to that of of Tyr +25 in the ^{15}N NMR spectrum of Pf1 coat protein.

DISCUSSION

The spectra in Figures 2 and 3 demonstrate two of the principal advantages of studying protein species prepared by solid-phase peptide synthesis. These spectra are on protein species that are difficult to handle, because of their hydrophobic character. None the less, it is possible to obtain high resolution spectra of similar quality on protein samples obtained by chemical synthesis and biosynthesis. The spectra in Figures 2 and 3 include spectra from species that can not be obtained from bacterial growth. The M13 procoat protein segment that is a substrate for leader peptidase (residues -15 to +8) in Figures 2A and 2B and the Pf1 coat protein (residues -36 to +46) are only available by chemical synthesis in quantities required for biophysical studies. The single line spectra in Figures 3A and 3B show that specific labelling is possible by chemical synthesis. In contrast, the best that can be hoped for from biosynthesis is selectively labelling with minimal scrambling of the label.

These initial results suggest that the promise of using chemically synthesized membrane bound proteins in NMR spectroscopy can be realized. Future studies will include solid-state NMR spectroscopy and the effects of single and multiple mutations of the proteins.

ACKNOWLEDGMENTS

We thank Wulf Carson and Bing Cai for technical assistance with peptide synthesis.

REFERENCES

1. Kent SBH (1988). Chemical synthesis of peptides and proteins. Ann Rev Biochem 57:957.
2. Verner K, Schatz G (1988). Protein translocation across membranes. Science 241:1307.
3. Markowski L (1984). Structural diversity in filamentous bacteriophages. In McPherson A (ed): "The Viruses," New York: Wiley, p. 203.
4. Wickner WT, Lodish HF (1985). Multiple mechanisms of protein insertion into and across membranes. Science 230:400.
5. Bogusky MJ, Leo GC, Opella SJ (1988). Comparison of the dynamics of the membrane-bound form of fd coat protein in micelles and in bilayers by solution and solid-state nitrogen-15 NMR spectroscopy. Proteins 4:123.
6. Opella SJ, Stewart PL, Valentine KG (1987). Protein structure by solid-state NMR spectroscopy. Q Rev Biophys 19:7.
7. Kuhn A, Wickner WT (1985). Isolation of mutants in M13 coat protein that affect its synthesis, processing, and assembly into phase. J Biol Chem 260:15907.
8. Tomich JM, Carson LW, Kanes KJ, Vogelaar NJ, Emerling MR, Richards JH (1988). Prevention of aggregation of synthetic membrane-spanning peptides by addition of detergent. Anal Biochem 174:197.
9. Pegg D, Doddrell D, Bendall J (1982). Distortionless enhancement of NMR signals by polarization transfer. J Mag Reson 48:323.
10. Dierstein R, Wickner W (1986). Requirements for substrate recognition by bacterial leader peptidase. EMBO J 5:427.
11. Rowitch DH, Perham RN (1987). Cloning and expression of tehe filamentous bacteriophage Pf1 major coat protein gene in *Escherichia coli*. J Mol Biol 188:873.

Frontiers of NMR in Molecular Biology, pages 119-127

THE HELIX-TURN-HELIX SUBDOMAINS OF LEXA AND LAC REPRESSORS[1]

R. Kaptein,[2] R.M.J.N. Lamerichs,[2] R. Boelens,[2] A. Padilla,[2] H. Rüterjans,[3] and M. Schnarr[4]

Department of Chemistry, University of Utrecht, Padualaan 8, 3584 CH Utrecht, The Netherlands

Institut für Biophysikalische Chemie, Johann Wolfgang Goethe-Universität, Theodor-Stern Kai 7, Haus 75A, D-6000 Frankfurt 70, Federal Republic of Germany

Institut de Biologie Moléculaire et Cellulaire du CNRS, Laboratoire de Biophysique, 15 rue René Descartes, 67084 Strasbourg Cédex, France

ABSTRACT A 2D NMR study of the N-terminal domain of LexA repressor shows that this protein has a helix-turn-helix domain in the peptide region 28-47, although its conformation differs from that of many other repressors. The spatial structure of the helix-turn-helix domain of LexA has been determined using the distance-geometry algorithm and is compared to that of lac, λ cI, and trp repressors.

---

[1] This work was supported by the Netherlands Foundation for Chemical Research (SON) with financial aid from the Netherlands Organisation for Scientific Research (NWO), by an INSERM grant (871 007) to M.G.-S. and by a research grant from the European Community (ST2J-0291).

[2] University of Utrecht.

[3] Johann Wolfgang Goethe-Universität, Frankfurt.

[4] Institut de Biologie Moléculaire et Cellulaire du CNRS, Strasbourg.

INTRODUCTION

Ever since the first X-ray structures of specific sequence DNA binding proteins, such as phage λ cro and cI repressors (1,2) and CAP (3), became available several research groups have tried to find general rules (or a "code") for protein-DNA recognition (for a review see ref. 4). Although recently the X-ray structures of repressor-operator co-crystals have been solved for trp repressor (5), λ cI repressor (6) and 434 repressor and cro (7,8), the prospects for such a recognition code have not become much better. These repressors all contain a helix-turn-helix (h-t-h) motif that carries most of the determinants for sequence specificity. However, the variety of the interactions used in DNA binding is so large that it would have been very difficult to predict them.

We have shown by NMR that the N-terminal DNA binding domain (headpiece) of the lac repressor also contains a h-t-h motif (9,10) and that this part of the protein interacts with lac operator (11). Here again a variation on the theme was seen in that the h-t-h domain in lac headpiece binds in an orientation with respect to the pseudo two-fold axis of lac operator, which is opposite to that so far observed for all other repressors.

Another interesting case is the LexA repressor, which binds to several operators of the SOS system of E.coli (12,13). The binding affinity to these operators varies in accordance with the kinetics required for the SOS response in the case of DNA damage. On the basis of sequence homology LexA has been considered a helix-turn-helix protein (4). However, interestingly the homology is so weak that using a recent algorithm devised to search for h-t-h domains LexA scores very low and moreover, a h-t-h motif is found starting at position 5 (in what turns out to be the wrong part of the protein!). Evidence from 2D NMR now shows that LexA repressor has a h-t-h structure starting at Arg 28. However, its conformation is significantly different from that found in the other repressors (14).

Here we discuss the NMR results obtained for lac headpiece and the N-terminal domain of LexA repressor and compare the spatial structure of the h-t-h domains of these proteins.

RESULTS AND DISCUSSION

Lac Repressor Headpiece.

Previous NMR studies (9,10) have shown that the N-terminal domain (51 residues) of lac repressor has three α-helices. The first two of these in the peptide region 6-25 constitute a h-t-h structure.

6				10					15
Leu -	Tyr -	Asp -	Val -	Ala -	Glu -	Tyr -	Ala -	Gly -	Val

α I (Leu 6 – Ala 13)

16				20					25
-Ser -	Tyr -	Gln -	Thr -	Val -	Ser -	Arg -	Val -	Val -	Asn

α II (Tyr 17 – Asn 25)

Sequence of lac repressor 6-25 and α-helices of the helix-turn-helix motif.

The α-helical secondary structure of lac headpiece is very well characterized by consecutive sets of strong sequential amide-amide NOEs implying short d_{NN} distances (15). Furthermore, extensive sets of NOEs corresponding to $d_{\alpha N}(i, i+3)$ and $d_{\alpha N}(i, i+4)$ are observed (9,16). Typical 1-4-5 patterns of hydrophobic residues occur in both helix I (Leu 6, Val 9, Ala 10) and helix II (Val 20, Val 23, Val 24).

The tertiary structure of lac headpiece was determined from a set of short proton-proton distance constraints from 169 NOEs (10,17). Using a combination of distance-geometry and restrained molecular dynamics a family of structures was generated, which satisfied the NOE constraints very well (largest violation 0.5 Å) and also had a low internal energy. The average runs difference between these structures was 0.8 Å for the backbone atoms of the residues 4-47. The N-terminal and C-terminal tripeptides are flexible resulting in a large conformational variability. The backbone structure of lac headpiece is shown in Figure 1.

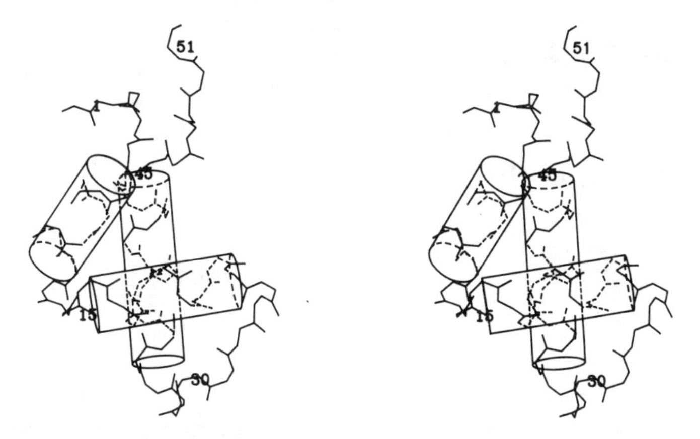

FIGURE 1. Stereo diagram of lac repressor headpiece backbone atoms. The cylinders represent α-helices at positions 6-13, 17-25, and 34-45.

The h-t-h motif of lac headpiece consisting of helices I and II adopts a folded structure, which is determined by several critical NOEs. Thus, strong NOEs are observed between the conserved residues Ala 10 and Val 20 (see Figure 2). Similar van der Waals contacts occur for the homologous residues in other repressors (4). The NOEs between Ala 13 and Val 15, also shown in Figure 2, are important in fixing the conformation of the turn between helices I and II. Furthermore, NOEs between residues of helices I and II include those of Leu 6 with Tyr 17 and Val 24, and of Tyr 17 with Tyr 7 and Ala 10 (16). Thus, in the spatial structure of lac headpiece the two α-helices of the h-t-h motif adopt a specific relative orientation, which is well determined by the NOEs and which , as we shall see below, is highly conserved among many DNA binding proteins.

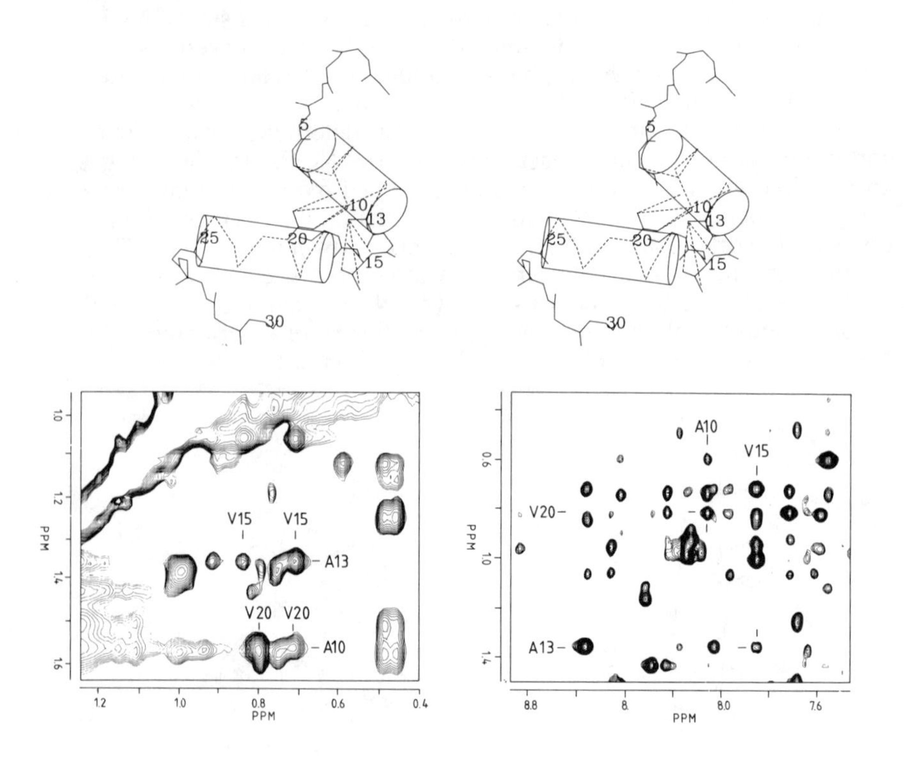

FIGURE 2. Parts of the 2D NOE spectrum of lac headpiece showing NOEs between Ala 10 and Val 20, and between Ala 13 and Val 15, that are critical for folding the helix-turn-helix domain. A stereo view of this domain is shown on top.

LexA Repressor.

Induction of the SOS genes of E.coli occurs through the action of RecA protein, which cleaves LexA repressor yielding an N-terminal domain of 84 amino acid residues. The decreased affinity for the SOS operators that results, induces transcription of the genes coding for repair enzymes. We have undertaken a structural study using 2D NMR of this N-terminal domain of LexA (14). Sequence specific ^{1}H resonance assignments were made for the first 60 residues and for a few isolated residues in the region 60-84. The short and medium range NOE information for the region 1-60 from which the secondary structure can be derived is shown in Figure 3. The presence of strong d_{NN} NOEs combined with sets of medium range $d_{\alpha N}$(i, i+3), $d_{\alpha\beta}$(i, i+3) and $d_{\alpha N}$(i, i+4) NOEs indicate the presence of three helical regions. The first two are regular α-helices in the regions 8-20 and 28-35. The third one consisting of residues 41-54 has a kink or overwound region near residues 48-49 as evidenced by a relatively strong $d_{\alpha\beta}$(i, i+3) NOE and a $d_{\beta N}$(i, i+2) connectivity.

The positions of the α-helices in LexA are compared with those of λ cI, lac, and trp repressors in Figure 4. The sequence alignment is as suggested by Pabo and Sauer (4). It can be seen that LexA indeed has helices in the region where other repressors have their h-t-h domains, although the second helix appears to start somewhat later. Furthermore, the first helix of LexA is homologous with similar ones in λ cI and trp repressors.

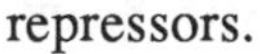

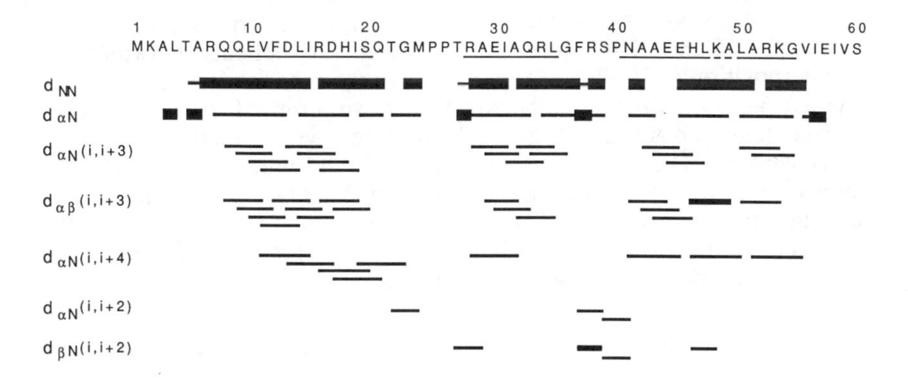

FIGURE 3. Short and medium range NOEs for the first 60 residues of the N-terminal domain of LexA repressor. Residues belonging to α-helical regions are underlined.

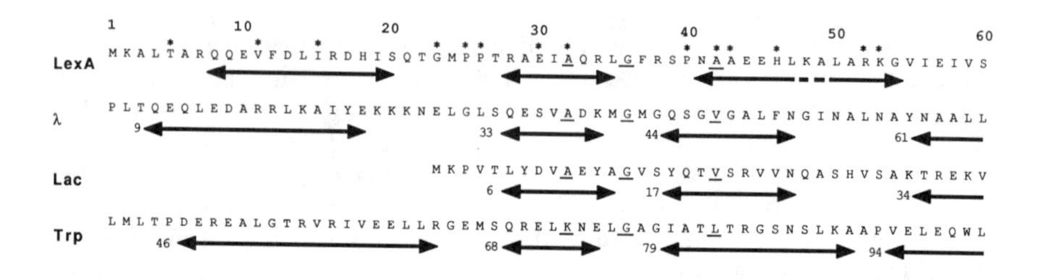

FIGURE 4. Positions of α-helical regions in LexA, λ cI, lac, and trp repressors.

Spatial Structure of the h-t-h Domain of LexA.

Inspection of the 2D NOE spectrum of LexA repressor shows that the NOE between Ala 32 and Ala 42, corresponding to the critical hydrophobic contact between Ala 10 and Val 20 in lac headpiece is missing. Instead, a large number of NOEs (twenty-three) is present involving Phe 37 suggesting that the side chain of this residue is positioned between helices 2 and 3 of LexA. The only direct NOE between these helices is the one between the C_α-proton of Arg 28 and the C_4 proton of His 46.

In order to determine the spatial folding of the h-t-h domain of LexA distance-geometry calculations were carried out for the peptide segment 28-47 using the program DISGEO (18). Distance constraints were based on a set of 28 NOEs observed between residues in this segment. The regions 28-35 and 41-47 were constrained in an α-helical conformation. A superposition of 10 distance-geometry structures is shown in Figure 5. The average rms difference between these structures for the backbone coordinates is 0.8 Å indicating that the folding is very well determined by the constraints. The structure shows indeed a bridging role for Phe 37 between the two helices of the h-t-h domain. This is shown more clearly in Figure 6. A network of van der Waals contacts involving Phe 37, Ile 31, Ala 32, Leu 35, Ala 42 and Ala 43 appears to determine the relative positioning of the two helices in this case.

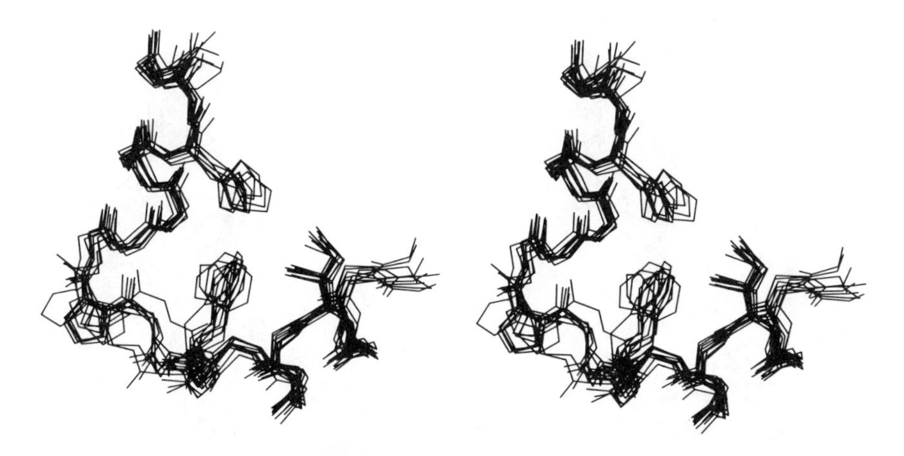

FIGURE 5. Overlay of 10 distance-geometry structures for the h-t-h region 28-47 of LexA repressor.

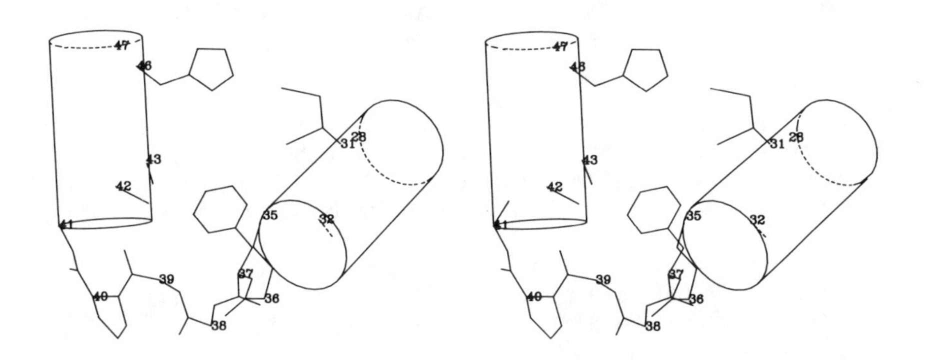

FIGURE 6. Stereo view of the h-t-h region of LexA repressor. A number of side drains are shown that show critical NOEs for folding of the domain.

A comparison between the h-t-h domain of LexA with those of lac and trp repressors and CAP is shown in Figure 7. For the alignment a best fit was calculated for the backbone atom positions of the first helix. The close correspondence in folding of the h-t-h domains of lac, trp and CAP is obvious from Figure 7 (average rms difference 0.8 Å). For LexA repressor the backbone conformation follows that of the others closely up to Gly 36. Then it starts to deviate for Phe 37 allowing its side chain to adopt a position between the helices. An extended peptide conformation is

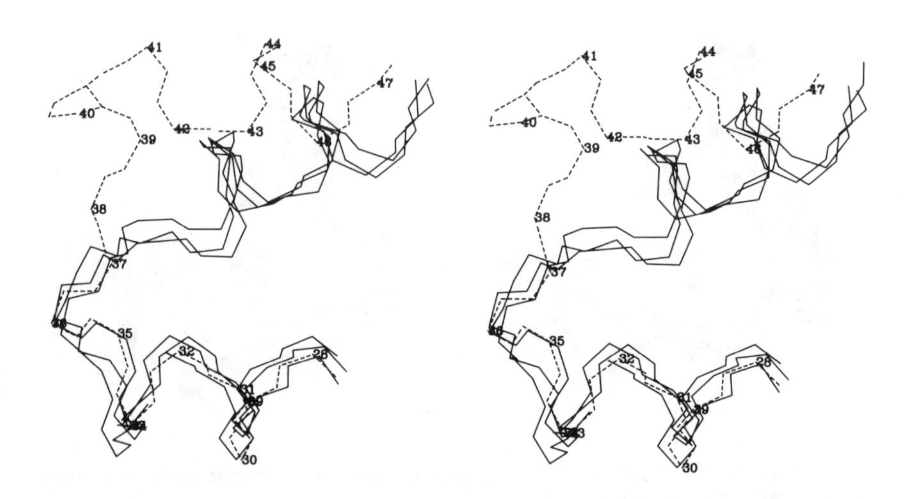

FIGURE 7. Overlay of the backbone conformation of the h-t-h domains of LexA, lac headpiece, trp repressor and CAP. The deviating structure (dashed line) is that of LexA repressor.

then present up to Pro 40. The net effect is that the second helix is displaced mainly at its amino-terminal end, while the relative orientation of the two helices can still be considered as rather similar to that of the other repressors. The h-t-h domain of LexA differs by 2.3 Å from that of the others.

The following conclusions can be made from this work:

- Residues 1-60 of LexA repressor contain three α-helices in the regions 8-20, 28-35 and 41-54. The last helix has a distortion (kink or overwound region) near residues 48-49.
- In spite of weak sequence homology with other repressors LexA has a helix-turn-helix motif in the region 28-47. The folding of this domain, however, is different from that of other DNA binding proteins.
- For lac repressor headpiece the relative positioning of the two helices is determined by a close contact (seen as a strong NOE) between residues Ala 10 and Val 20. In the case of LexA a similar role is played by a network of van der Waals contacts involving Phe 37, the side chain of which is positioned between the two helices.

REFERENCES

1. Anderson WF, Ohlendorf DH, Takeda Y, Matthews BW (1981). Nature (London) 290: 754.
2. Pabo CO, Lewis M (1982). Nature (London) 298: 443.
3. McKay DB, Steitz TA (1981). Nature (London) 290: 744.
4. Pabo CO, Sauer RT (1984). Ann Rev Biochem 53: 293.
5. Otwinowski Z, Schevitz RW, Zhang R-G, Lawson CL, Joachimiak A, Marmorstein RQ, Luisi BF, Sigler PB (1988). Nature (London) 335: 321.
6. Jordan SR, Pabo CO (1988). Science 242: 893.
7. Anderson JE, Ptashne M, Harrison SC (1987). Nature (London) 326: 846.
8. Wolberger C, Dong Y, Ptashne M, Harrison SC (1988). Nature (London) 335: 789.
9. Zuiderweg ERP, Kaptein R, Wüthrich K (1983). Proc Natl Acad Sci USA 80: 5837.
10. Kaptein R, Zuiderweg ERP, Scheek RM, Boelens R, van Gunsteren WF (1985) J Mol Biol 182: 179.
11. Boelens R, Scheek RM, van Boom JH, Kaptein R (1987). J Mol Biol 193: 179.
12. Little JW, Mount DW (1982). Cell 29: 11.
13. Walker GC (1984). Microbiol Rev 48: 60.
14. Lamerichs RMJN, Padilla A, Boelens R, Kaptein R, Ottleben G, Rüterjans H, Granger-Schnarr M, Oertel P, Schnarr M (1989). To be published.
15. Wüthrich K (1986). NMR of proteins and nucleic acids (Wiley, New York).
16. Zuiderweg ERP, Scheek RM, Kaptein R (1985). Biopolymers 24: 2257.
17. De Vlieg J, Boelens R, Scheek RM, Kaptein R, van Gunsteren WF (1986). Israel J Chem 27: 181.
18. Havel TF, Wüthrich K (1984). Bull Math Biol 45: 673.

Frontiers of NMR in Molecular Biology, pages 129-143

STATIC AND DYNAMIC ASPECTS OF PROTEIN STRUCTURE[1]

G. Wagner, N.R. Nirmala, G.T. Montelione, and S. Hyberts

Biophysics Research Division, University of Michigan,
Ann Arbor, Michigan 48109

ABSTRACT We have developed novel techniques to characterize by NMR static and dynamic aspects of protein structure. These include a Taylor transformation of NOESY mixing time series which allows a convenient determination of cross relaxation rates and an identification of spin diffusion effects. Novel pulse sequences designed for ^{15}N labeled polypeptides allowed accurate measurements of homonuclear and long-range heteronuclear coupling constants. Furthermore, we have measured T_1 and T_2 relaxation times of individually assigned carbon resonances in a protein, using 2D heteronuclear COSY experiments. With related experiments we have identified and characterized a multiple conformation in the reactive site of a protease inhibitor.

INTRODUCTION

Most proteins function by adapting a well defined three-dimensional conformation. It is not yet clear, however, to which detail the structure has to be known to make functional

---

[1]This work was supported by NSF (Grants DMB-8616059 and BBS-8615223), NIH (Grant GM 38608), and the Damon Runyon - Walter Winchell Cancer Research Fund (Grant DRG-920).

statements, and whether and to what degree protein dynamics and stability are relevant for protein function. Certainly this depends on the kind of function a particular protein has. At present, it is possible to check relationships between protein function and structural aspects by replacing single amino acids and following structural and functional changes. It appears that, even minor structural changes can entirely alter the function of a protein. Therefore, very accurate structure determinations have to be carried out to make relevant statements about the structural basis of protein function. We have made attempts to improve our knowledge about proteins by developing new approaches to characterize static and dynamic aspects of proteins.

TAYLOR TRANSFORMATION OF NOESY τ_m SERIES

In the past, theories and strategies have been developed to determine proton-proton distances from NOESY experiments. The mixing time dependence of cross peak intensities can be described as (1):

(eqn1) $a_{kl}(\tau_m)=[\exp(-\mathbf{R}\ \tau_m)]\ a_{ll}(0)$

This can be developed in a Taylor series:

(eqn2) $a_{kl}(\tau_m) \approx (\delta_{kl}-R_{kl}\tau_m+1/2\Sigma_j\ R_{kj}R_{jl}\tau_m^2 + ..)$

For macromolecules, such as proteins we have:

(eqn3) $R_{kl} \geq 0$ for $k=l$
$R_{kl} \leq 0$ for $k \neq l$

To simplify the data analysis, we have developed computer software to fit entire NOESY spectra of a mixing time series, point by point, to a Taylor series, prior to peak integration (2). By this procedure we transform the spectra of a NOESY mixing times series into 2D spectra of Taylor coefficients. Ideally, the zero-order spectrum would have no intensity for cross peaks. The cross peak intensities in the first-order spectrum are directly proportional to the cross relaxation rate. Peak integration in this first-order spectrum provides a reliable measure for cross relaxation rates without the need to integrate cross peaks in several NOESY spectra acquired with different mixing time. The second-order spectrum contains contributions from spin diffusion and relaxation. The terms $R_{kj}R_{jl}$ in the sum of eqn. 2 can be positive or negative. The terms

(eqn4) $R_{kj}R_{jl} \geq 0$ for $j \neq k$ and $j \neq l$

represent spin diffusion, and the terms

(eqn.5) $R_{kj}R_{jl} \leq 0$ for $j=k$ or $j=l$

are due to relaxation and cross relaxation. Both contributions (4) and (5) tend to compensate each other.The second-order spectrum can thus be used for diagnostic purposes.Strong spin diffusion would result in positive peaks, otherwise the peaks are negative. We have recorded a series of 11 NOESY spectra with mixing times from 25 ms to 75 ms for the protein eglin c. This series was subject to the Taylor transformation as described above. We have truncated the series after the second order term. The zero-order term was set to zero. In this approximation, all cross peaks of the second order spectrum were negative. This means that in the mixing time range up to 75 ms spin diffusion is not dominant. In particular, we have analyzed NOE's to methylene protons (see Fig. 1). In the case considered, the NOE's were quite different in the 25 ms spectrum. At 75 ms they were almost equal in intensity. In the spectrum of the second-order Taylor coefficients the cross peak that is intense in the 25 ms spectrum is intense and negative whereas the other cross peak which corresponds to the more distant methylene proton is close to zero. This means that the nearly equal intensities of the two cross peaks to methylene protons is not due

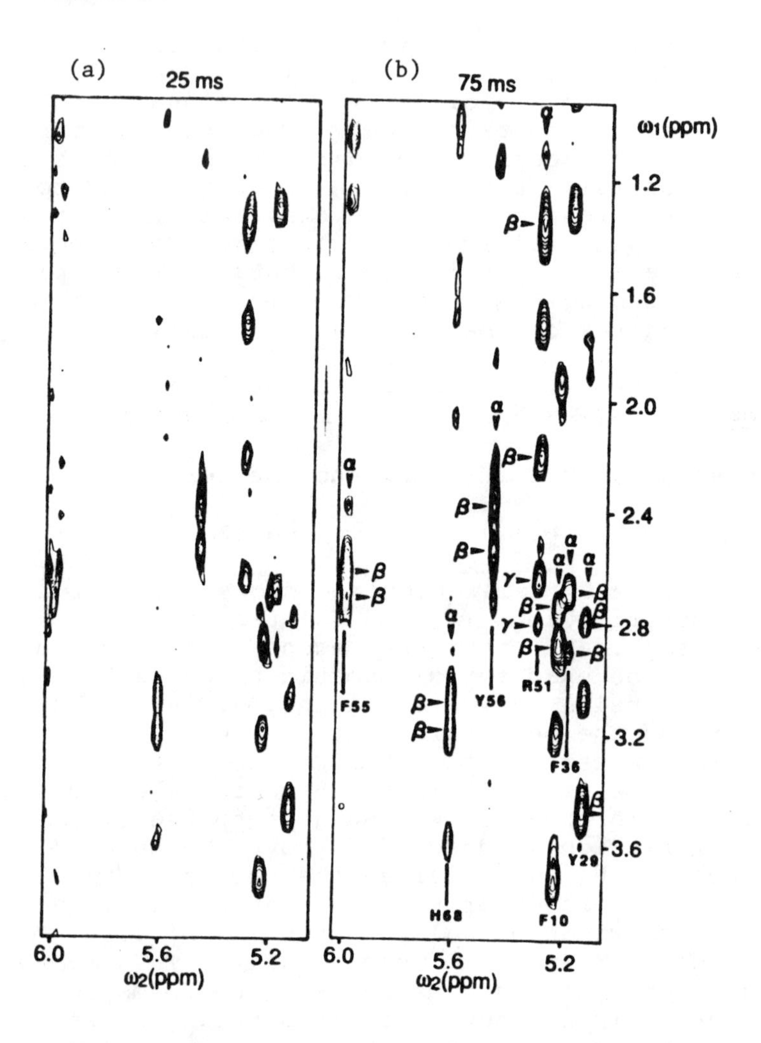

FIGURE 1. NOESY cross peaks between α-protons and side chain protons in the protein eglin c, recorded with mixing times of 25 ms (a) and 75 ms (b). The spectra of the first and second order Taylor coefficients are shown in (c) and (d), respectively.

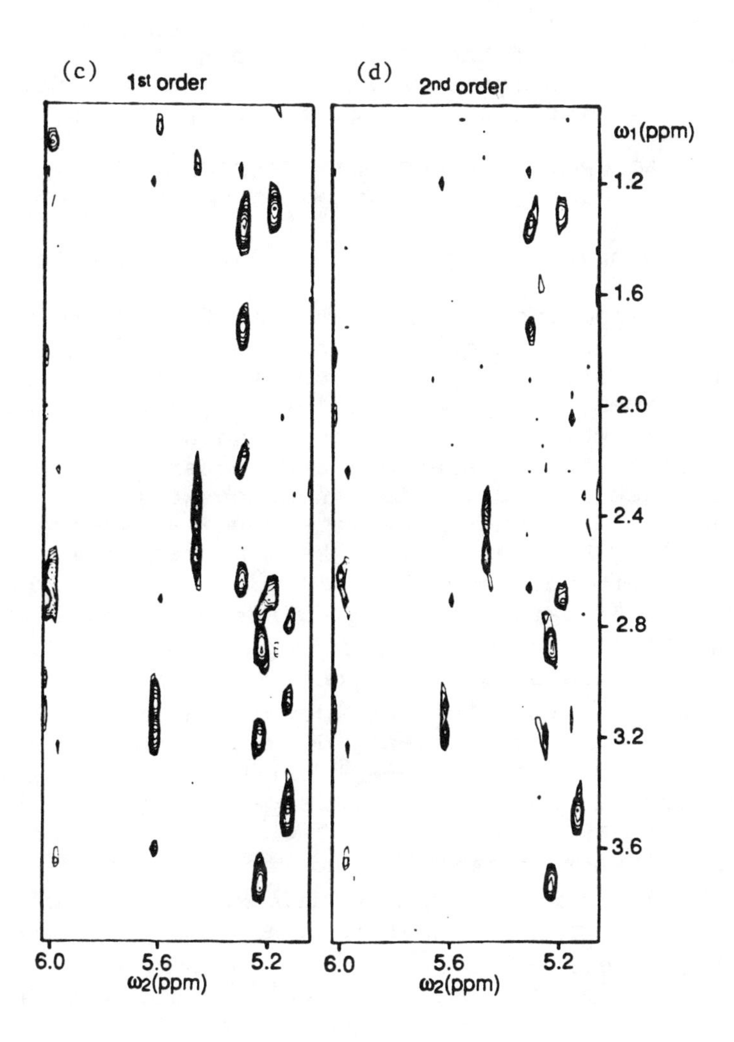

FIGURE 1 (continued).

to a steeper (concave) growth of the crosspeak receiving spin diffusion. It is rather due to a retardation of the effect of spin lattice relaxation which otherwise decreases and turns negative the slope of NOE build-up curves.

ACCURATE MEASUREMENTS OF COUPLING CONSTANTS IN ^{15}N LABELED POLYPEPTIDES

Heteronuclear Long-Range Coupling Constants

Stereospecific assignments of β-methylene protons is essential for obtaining high quality protein structures. Previously, stereospecific assignments have been achieved by analysis of vicinal coupling constants between α- and β-protons and intraresidue and sequential NOE's (3) (see Fig. 2). This situation would be improved if long-range heteronuclear vicinal coupling constants could be measured between the amide nitrogen and the β-protons. This coupling constants varies between -0 and -9 Hz (4). If the H^{α}-H^{β} coupling constants are both small, which means a gauche-gauche conformation with $\chi^1 = 60°$, $H^{\beta 2}$ and $H^{\beta 3}$ can be identified from large and small heteronuclear ^{15}N couplings, respectively.If the H^{α}-H^{β} coupling constants are one small and one large, respectively, (gauche-trans), a situation with two small ^{15}N couplings indicates a χ^1 of 180°, and the H^{β} gauche to H^{α} is $H^{\beta 2}$, whereas a situation with one large and one small ^{15}N coupling indicates a χ^1 of -60°, and the H^{β} gauche to H^{α} is $H^{\beta 3}$. This is widely known. However, it has been difficult to measure the small heteronuclear coupling constants accurately. Attempts to measure such heteronuclear coupling constants in proteins, for example from long-range heteronuclear correlation experiments had little success. Recently we have realized (5) that heteronuclear long-range coupling constants between the amide nitrogen and β-protons can easily and accurately be measured from H^N-H^{β} cross peaks in homonuclear

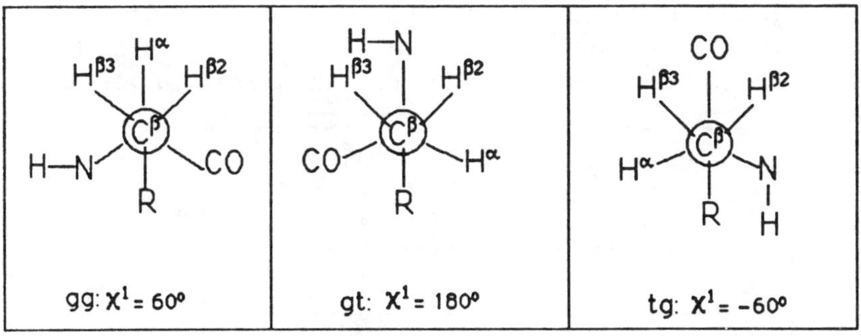

FIGURE 2. The three common staggered conformers of amino acid side chains containing β-methylene groups, corresponding to digedral angles χ^1 of 60°, 180°, and -60°, respectively. The three conformers can readily be distinguished by measuring the homonuclear and heteronuclear coupling constants from H^{α} and N to $H^{\beta 2}$ and $H^{\beta 3}$.

2D NMR experiments of ^{15}N labeled proteins where the nitrogen does not experience a mixing pulse between evolution and mixing period. In an ^{15}N-labeled protein, simple NOESY, RELAY or TOCSY experiments show the long range coupling in the H^N-H^{β} cross peaks (see Fig. 3). The cross peaks below the diagonal, at $\omega_1 = \omega(H^N)$ and $\omega_2 = \omega(H^{\beta})$ are most suited for these measurements. Each cross peak consists only of two components, located at:

$$[\omega_1, \omega_2] = [\omega(H^N) + {}^1J(N,H^N),\ \omega(H^{\beta}) + {}^3J(N,H^{\beta})]$$

$$[\omega_1, \omega_2] = [\omega(H^N) - {}^1J(N,H^N),\ \omega(H^{\beta}) - {}^3J(N,H^{\beta})]$$

Since the one-bond coupling $^1J(N,H^N)$ is very large there is no overlap between the two components and the heteronuclear coupling constants $^3J(N,H^{\beta})$ can be measured with the same accuracy as chemical shifts. This accuracy can even be improved by determining the first moments (center of mass) of the components along ω_2. Obviously, these experiments can also be used to measure coupling constants between the amide nitrogen of residue i and the α-proton of residue i-1.This provides access to the dihedral angle ψ.

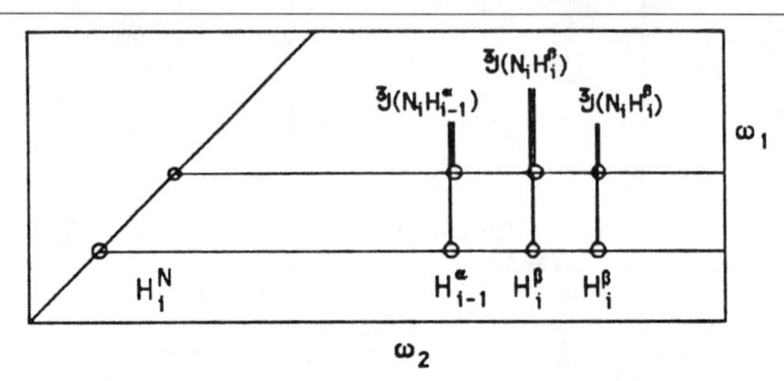

FIGURE 3. Schematic representation of NOESY cross peaks of amide protons of residue i to α-protons and β-protons of the same residue i, and α-protons of residue i-1, in uniformly ^{15}N enriched proteins.

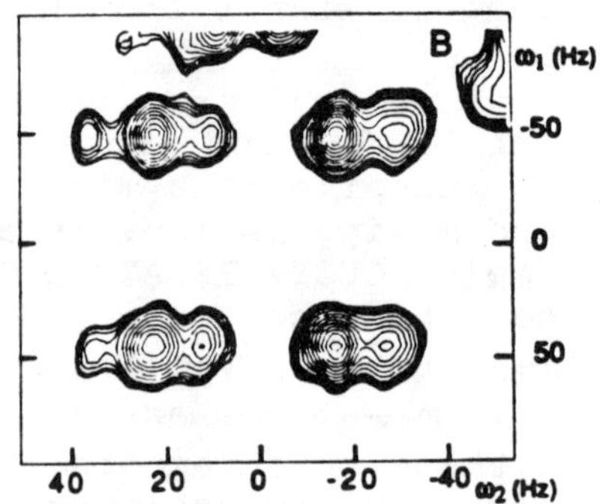

FIGURE 4. H^N-$H^β$ NOESY cross peaks for the residue Cys 21 of uniformly ^{15}N labeled human TGFα.The long range heteronuclear couplings $^3J(N$-$H^β)$ are -1.3 and -1.8 Hz, for $H^{β2}$ and $H^{β3}$, respectively.

We have examined this approach on uniformly ^{15}N labeled human TGFα. Some cross peaks are shown in Fig. 4. ^{15}N labeling of the protein to be studied is very useful but not absolutely necessary for these experiments. If the homonuclear proton-proton 2D NMR experiment (NOESY, RELAY, TOCSY) is preceded by a BIRD or TANGO sequence, the same information can be obtained at natural abundance ^{15}N, however at much lower sensitivity. The most serious problem for application of this technique is the overlap of the many (H^N,H^β) cross peaks. This can be overcome by expanding the sequence to a 3D experiment with a ^{15}N evolution period prior to the experiment described.

Homonuclear vicinal coupling constants $^3J(H^\alpha-H^N)$

Obviously, the concept of the experiments described above can be extended for measurements of homonuclear coupling constants $^3J(H^\alpha-H^N)$. We have designed 2D $^1H-^{13}C-^{15}N$ triple resonance experiments (6) that provide, either cross peaks between H^N and C^α where H^α is not effectively pulsed between evolution and detection, or cross peaks between H^α and N where H^N is not effectively pulsed between evolution and detection.We have carried out the former experiment.To achieve a decent signal-to-noise ratio it requires isotope enrichement of the peptide nitrogen or the α-carbon with ^{15}N or ^{13}C, respectively. The pulse sequence (6) is shown in Fig. 5. The experiment has been tested on the tripeptide Ac-Asn-Pro-(^{15}N)Tyr-NHMe. A schematic of the resulting cross peak is presented in Fig. 6.The experiment consists of a polarization transfer from H^α to C^α by a refocussed INEPT sequence. The carbon coherence is frequency labeled during t_1, decoupled from nitrogen but coupled to proton. This produces two multiplet components (along ω_1) which correspond to two ensembles of molecules which have the H^α spin either up or down, respectively. Both coherence components are then transferred to nitrogen by a INEPT-like sequence. During this process the coherence is decoupled from protons. Now we want to transfer the coherence from the peptide nitrogen to H^N without mixing the two coherence components evolved during t_1. This is achieved via a refocussed INEPT transfer to H^N where the first proton pulse is replaced with a TANGO (7) pulse, which is a 90°-pulse selective for H^N and does not affect H^α. As a consequence, during t_2, all H^N coherence that experiences the coupling to a spin up H^α had seen the same H^α spin up already during t_1, and vice versa. Therefore, we see only two cross-peak components with the coordinates:

$$[\omega_1,\omega_2] = [\omega(C^\alpha)+{}^1J(C^\alpha,H^\alpha),\ \omega(H^N)+{}^3J(H^N,H^\alpha)]$$
$$[\omega_1,\omega_2] = [\omega(C^\alpha)-{}^1J(C^\alpha,H^\alpha),\ \omega(H^N)-{}^3J(H^N,H^\alpha)].$$

Various 3D versions of the experiments described in this section are currently being developed in our laboratory to avoid severe overlap of cross peaks. Alternative 2D experiments are also being investigated and will be described elsewhere.

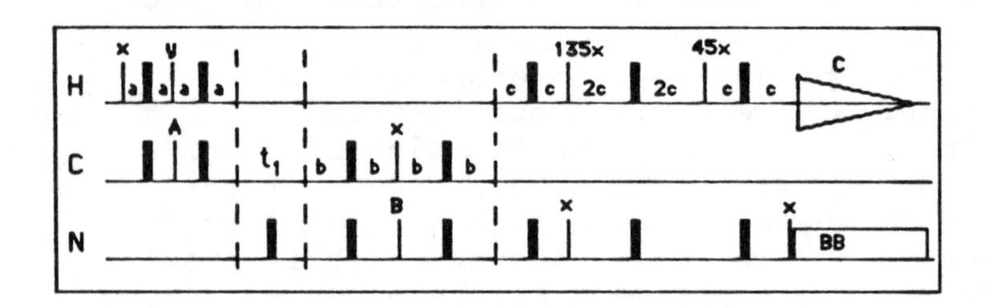

FIGURE 5. Pulse sequence of a 2D triple resonance experiment for measurements of homonuclear vicinal coupling constants between amide protons and α-protons.

^{13}C RELAXATION TIMES

^{13}C relaxation time measurements can be used to study fast reorientational motions of C-H bond vectors. 1D experiments for measurements of ^{13}C relaxation times do not provide sufficient resolution to follow the relaxation of

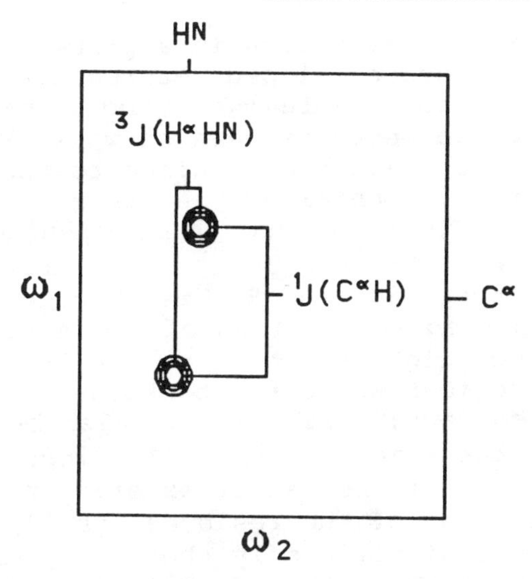

FIGURE 6. Schematic drawing of a C^{α}-H^{N} cross peak obtained with the experiment of Fig. 5.

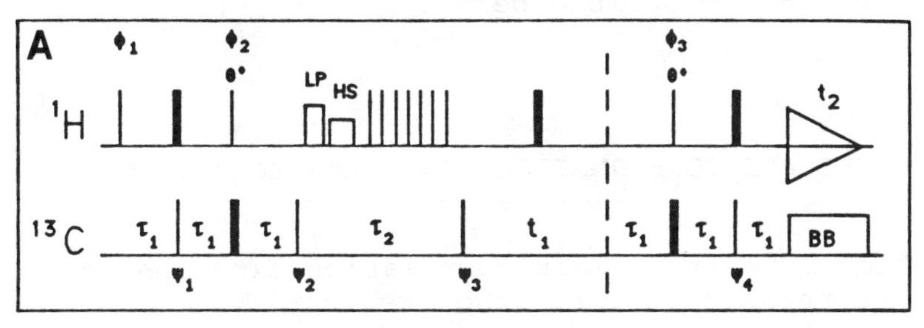

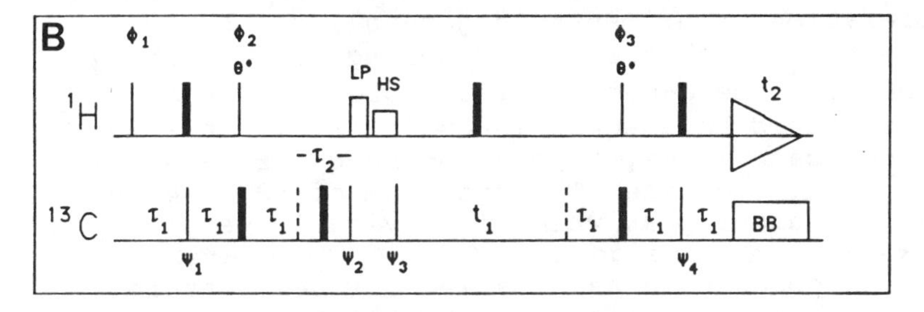

FIGURE 7. Pulse sequences for measurements of ^{13}C T_1 (A) and T_2 (B) values in 2D heteronuclear correlation experiments.

individual carbon resonances in a protein. For this reason we have developed heteronuclear 2D NMR experiments to measure ^{13}C T_1 and T_2 values (8,9).These pulse sequences are shown in Fig. 7. These experiments have been applied to the basic pancreatic trypsin inhibitor, and an almost complete set of T_1-values has been measured for the α-carbons, at pH 4.5, 36°C. The relaxation times vary between 300 and 500 ms. The longest T_1's were measured for regions of the molecule where we expect highest mobility, the C-terminal Ala 58, and regions where the backbone reaches the protein surface.T_2 values have also been measured for the same α-carbons. The most striking feature of the latter experiments was that the α-carbons of the residues Cys 14, Cys 38 and Tyr 35 have significantly shorter T_2 values than the rest of the molecule. This is indicative of an intramolecular exchange process which will be described below.

MULTIPLE CONFORMATIONS IN THE REACTIVE SITE OF BPTI

The 2D heteronuclear correlation experiments, such as those used for carbon relaxation experiments (8,9), and one-bond carbon-proton correlation experiments in general, can be tuned perfectly to the one-bond coupling between the α-carbon and the α-proton. Therefore, all cross peaks should have equal intensities. If this is not the case we may have an exchange process broadening the line. The spectrum of BPTI shows one such cross peak which is obviously very broad at 36°C, assigned to Cys 14. To clarify this phenomenon we have varied the temperature. Fig. 8 shows expansions of the spectra around this cross peak at different temperatures. The peak is intense at low temperature, it broadens

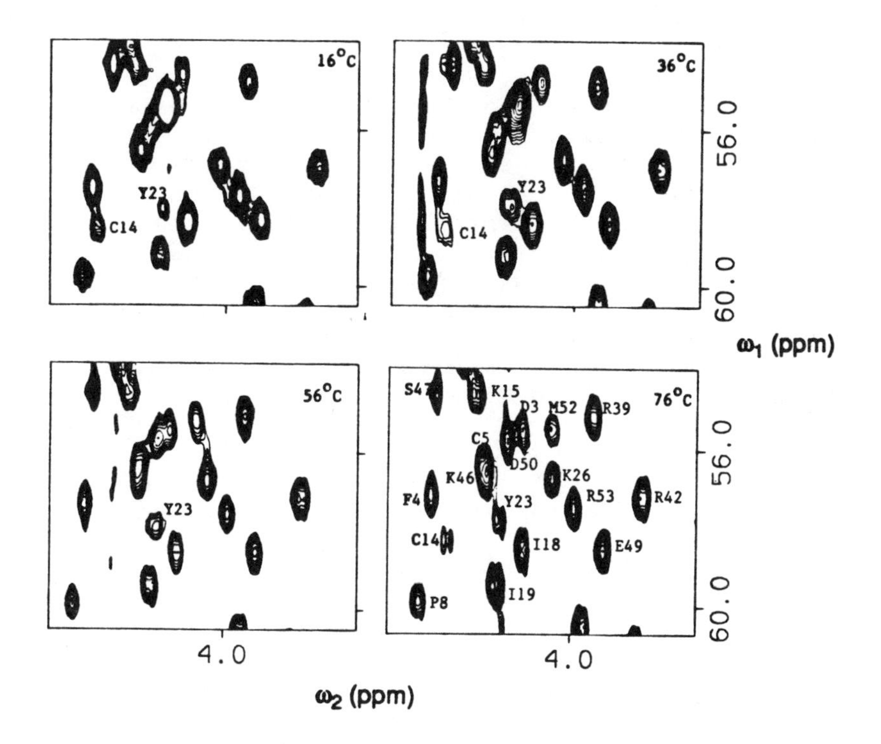

FIGURE 8. Temperature dependence of heteronuclear $C^{\alpha}H$ COSY cross peaks of some residues of BPTI, including Cys 14.

and disappears above around 36°C, and it reappears at temperatures above 60°C. The positions at 6°C and 76°C differ by ca. 40 Hz and 25 Hz in the carbon and proton dimension, respectively. Cys 14 is preceded by Pro 13. We were suspicious whether this dynamic phenomenon could be related to cis-trans isomerization of this proline residue. However, no temperature dependent spectral features could be seen on this residue which would suggest such a relation. On the other hand, the short T_2 of Cys 38 indicates that the whole disulfide bond 14-38 is involved in the dynamic process. It seems possible that the disulfide bond switches between two different conformations with variation of the temperature. Such disulfide dynamics may be a common feature in proteins, probably of similar importance to protein dynamics as proline cis-trans isomerization.

ACKNOWLEDGEMENTS

We thank Dr. M. Winkler and Genentech, Inc. for providing ^{15}N labeled TGFα, and Dr. Y.C. Meinwald for providing Ac-Asn-Pro-[^{15}N]Tyr-NHMe.

REFERENCES

1. Macura S, Ernst, RR (1980) Mol Phys
2. Hyberts S, Wagner G (1989) Taylor transformation of 2D NMR τ_m series from time dimension to polynomial dimension. J Magn Reson, in press.
3. Hyberts S, Märki W, Wagner G (1987) Stereospecific assignments of side chain protons and characterization of torsion angles. Eur J Biochem, 164:625.
4. Bystrov V (1976) Spin-spin coupling and the conformational states of peptide systems. Progr NMR Spectroscopy, 10:41.
5. Montelione GT, Winkler ME, Rauenbuehler P, Wagner G (1989) Accurate measurements of long-range heteronuclear coupling constants from homonuclear 2D NMR spectra of isotope-enriched proteins. J Magn Reson, in press.

6. Montelione GT, Wagner G (1989) Accurate Measurements of Homonuclear H^N-H^a Coupling Constants in Polypeptides Using Heteronuclear 2D NMR Experiments. J Am Chem Soc, submitted.
7. Wimperis S, Freeman R (1984) An Excitation Sequence which discriminates between direct and long-range CH coupling. J Magn Reson, 58:348.
8. Nirmala NR, Wagner G (1988) Measurements of ^{13}C relaxation times in proteins by two-dimensional heteronuclear ^1H-^{13}C correlation spectroscopy, J Am Chem Soc, 110:7557.
9. Nirmala NR, Wagner G (1989) Measurement of ^{13}C spin-spin relaxation times by two-dimensional heteronuclear ^1H-^{13}C correlation spectroscopy, J Magn Reson, in press.

Frontiers of NMR in Molecular Biology, pages 145-153

Structural Differences of Transforming *ras* p21(Val-12) From the Normal Protein[1]

L. Tong, A.M. de Vos, M.V. Milburn, J. Jancarik, S. Noguchi[2], S. Nishimura[2], K. Miura[3], E.Ohtsuka[3] and S.-H. Kim[4]

Department of Chemistry and Lawrence Berkeley Laboratory, University of California, Berkeley, CA 94720 U.S.A.

ABSTRACT One of the most commonly found transforming *ras* oncogenes in human tumors has a valine codon replacing the glycine codon at position 12 (1). To obtain structural basis for understanding cell transformation by this single amino acid substitution, we have determined the crystal structure of the GDP bound form of this mutant, p21(Val-12). One of the major differences between this structure and that of the normal protein (2) is that the loop that binds the ß-phosphate of GDP is enlarged. Such a change in the "catalytic site" conformation could explain the reduced GTPase activity of the mutant (3), and presumably keeps the protein in the GTP bound "signal on" state for a prolonged period of time, ultimately causing cell transformation. We describe the overall structure of p21(Val-12) at 2.2Å resolution and discuss in detail one of the major differences from the structure of the normal c-H-*ras* protein (2). Structural comparison of normal and transforming *ras* proteins provides a basis for understanding cell transformation at the molecular and structural level.

---

[1]This work was supported by grants from the National Institute of Health, the Department of Energy, and the Japanese Ministry of Health and Welfare.

[2]Biology Division, National Cancer Center Research Institute, Tokyo, Japan.

[3]Faculty of Pharmaceutical Sciences, Hokkaido University, Sapporo, Japan.

[4]To whom correspondence should be addressed.

INTRODUCTION

The *ras* oncogene is one of the most comonly found oncogenes in human cancer cells and the gene products, *ras* oncoproteins, are thought to belong to a unique class of oncogene, the signal transducer. The most widely accepted model of the mechanism of cell transformation by *ras* proteins is that the external signal perceived by the receptor for cell growth or differentiation is transmitted to the *ras* proteins which in turn initiate the first step in the signal cascading inside the cell. The *ras* complexed with GTP is considered to be a "signal-on" state and that complexed with GDP a "signal-off" state. Normal *ras* proteins have an intrinsic GTPase activity. However, most of the transforming *ras* oncoproteins have been found to have either reduced or no GTPase activity, so that the transforming *ras* oncoproteins stay in a "signal-on" state which stimulates cell proliferation. A simplified version of our current view of how *ras* proteins work is shown in Fig. 1.

The difference between the normal and transforming *ras* oncogenes usually is a single point mutation which results in single amino acid substitution in one of a few key regions of the gene product, *ras* oncoprotein or p21. The recent determination of the crystal structure of human normal c-H-*ras* oncoprotein revealed that these key regions are all clustered around the GDP binding site.

One of the most common transforming mutations found in human tumors is a substitution of Gly at position 12 by a Val, which results in the drastic reduction in the GTP hydrolysis rate. To understand the critical nature of this single amino acid substitution, we have determined the crystal structure of the tranforming mutant, p21 Val-12 lacking the C-terminal 18 residues that are thought to be flexible. We report here the crystal structure of the protein at 2.2Å resolution and compare it with the normal c-H-*ras* protein also refined at 2.2Å resolution. Specifically we compare the "catalytic" site of the GTPase activity between these two proteins, where one of the major differences is found.

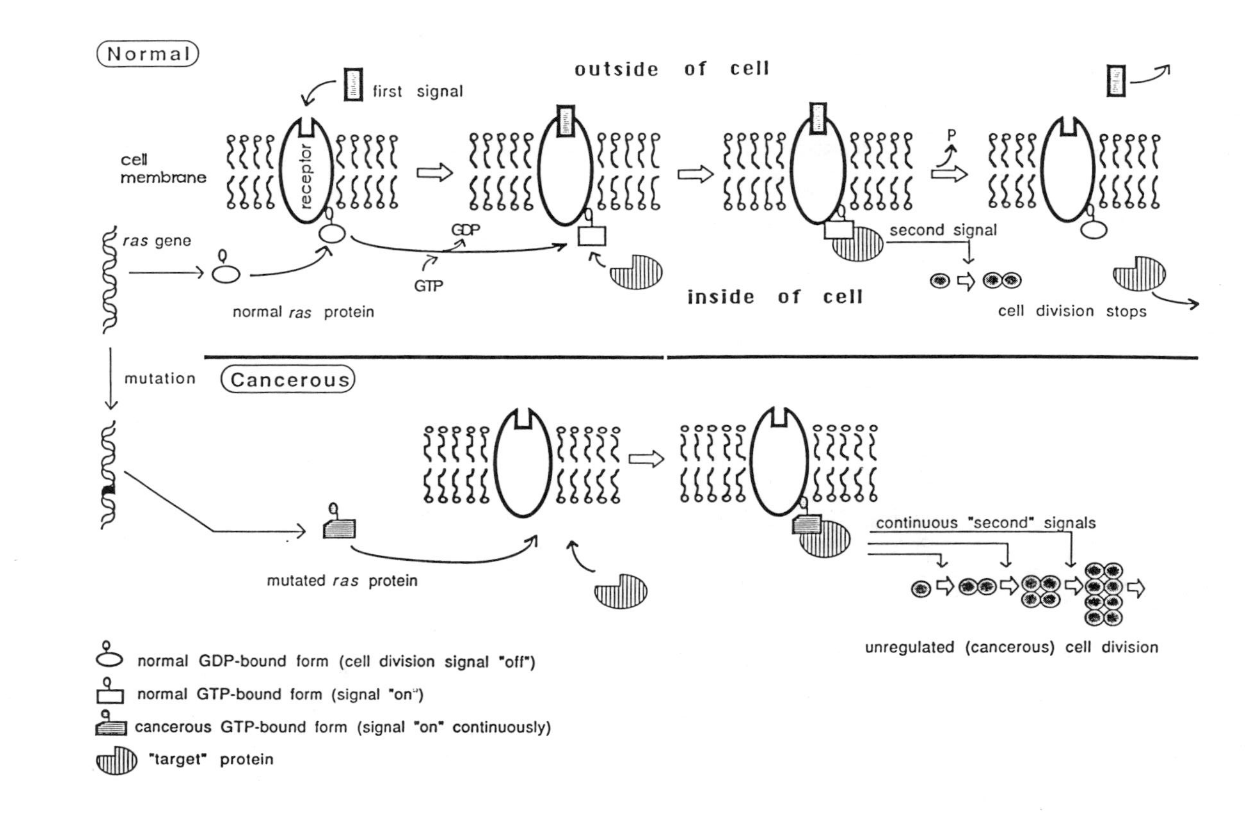

FIGURE 1. A simplified current view on the mode of action of *ras* oncoprotein.

METHODS

The cloning and expression (4) of the synthetic gene coding for human c-H-*ras* p21 proteins as well as the crystallization conditions (5) have been described. Briefly, the synthetic genes (using *E. coli* preferred codons) coding for amino acid residues 1-189 and 1-171 were cloned and expressed in *E. coli*. For the latter, we have deleted the codons corresponding to the C-terminal 18 residues because there is reasonable evidence to suspect that the region is flexible, and that its flexibility may be detrimental in obtaining good quality crystals. Only the soluble fraction of the protein was purified for crystallization. The GTPase activity of p21(Val-12) was less than 10% of that of the normal protein. High quality single crystals were obtained by vapor phase equilibrium technique from solutions containing 10 mg/ml protein, 0.1 M $CaCl_2$, 0.075 M Hepes buffer at pH 7.5, 0.5 mM EDTA, 0.5 mM DTT and 0.005% n-octyl glucoside equilibrated to 30% PEG 400. The crystals were isomorphous to those of the normal protein, and have space group $P6_522$ with cell parameters $a = b = 83.2$Å and $c = 105.1$Å.

The diffraction data for p21(Val-12, 1-171) was collected on an Enraf Nonius rotation camera installed on the 8-pole wiggler line at the Stanford Synchrotron Radiation Laboratory, Palo Alto, California. The x-ray wavelength used for data collection was 1.08Å, the crystal-to-film distance was 85mm, and 2° rotation was used for each exposure. One set of data was collected from one crystal, at 4°C, on a total of 32 films (16 film packs). A total of 48073 diffraction data were used in the merging process, giving 10447 unique reflections to 2.2Å resolution, with an overall R(merge) on intensity of 8%. Structure was determined by difference fourier method, calculated using phases obtained from the 2.7Å resolution structure of normal c-H-*ras* proteins. The largest single difference in electron density map was located near the residue position 12, as expected. Throughout the entire unit cell, there were no other comparably large electron density peaks visible.

During the course of crystallographic refinement, omit maps were calculated by removing residues 9-18 and 55-64, corresponding to loop L1 and part of L4, where most difference electron densities appear (Fig. 2), and the model was refitted to the omit maps. The current R factor

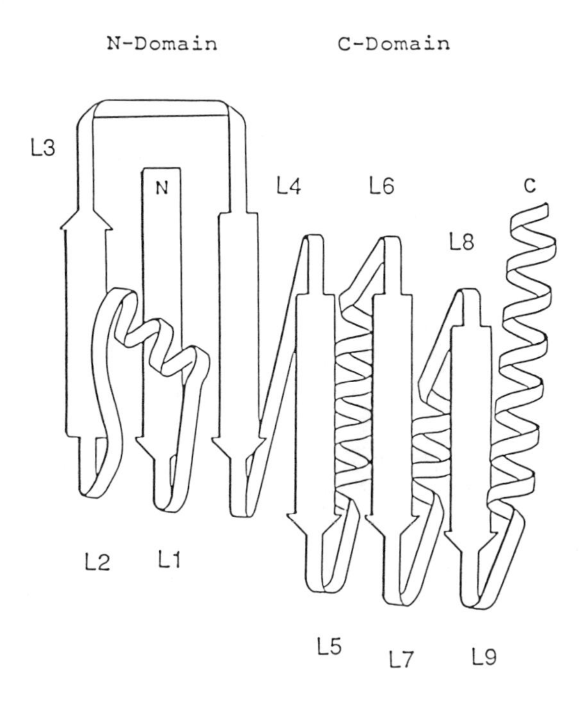

FIGURE 2. "Topological structure" of human c-H-*ras* p21 protein. Each ß-strand, α-helix and loop is numbered starting from the N-terminus. The N-domain, which covers three ß-strands and one α-helix, is barely hydrogen-bonded to the C-domain. The former has much higher thermal motion than the latter, suggesting that the flexibility of the N-domain is greater. The catalytic site for GTP hydrolysis is localized in the N-domain ("phosphate binding domain") and the recognition for guanine base is in the C-domain (guanine recognition domain).

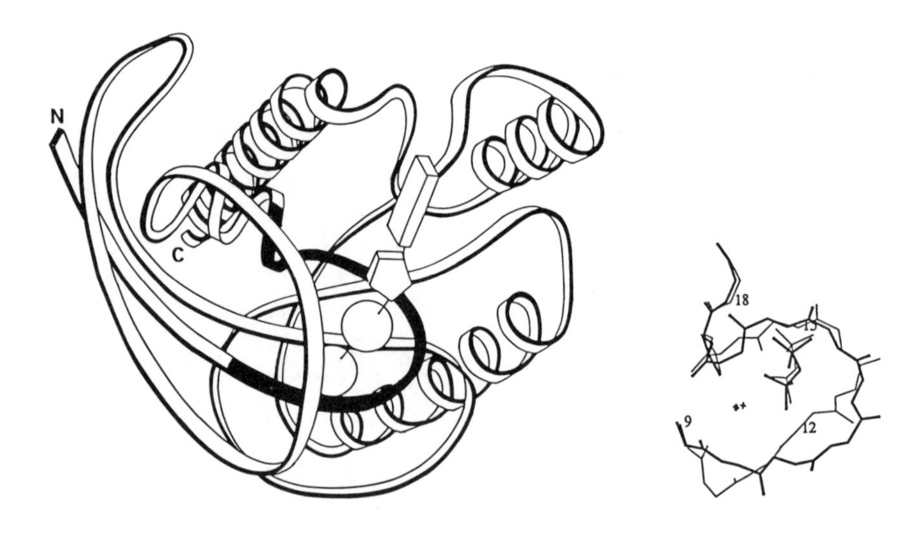

FIGURE 3. Overall architecture of human c-H-*ras* oncogene protein. The major structural difference between the normal and one of the most common cancerous *ras* proteins, p21(val-12), is confined in a looped region in black. Atomic details of the region are shown on the right, where thicker lines represent structure of the cancerous *ras* protein, p21(val-12), and thinner lines that of the normal *ras* protein.

for 9053 reflections (80% of total possible) between 10-2.2Å resolution is 23.6%.

RESULTS AND DISCUSSION

The overall structure (Fig. 3) of the transforming p21(Val-12) is similar to that of the normal protein (2). It contains six ß-strands, four α-helices, and nine connecting loops. Both structures appear to consist of two recognizable domains (Fig. 2): the N-terminal domain, containing the first 75 residues (including the first three ß-strands and one α-helix), is the "phosphate hydrolysis domain", and the C-terminal domain, containing the remaining residues (including the last three ß-strands and three α-helices), is the "guanine recognition domain". There is only a short stretch of hydrogen-bonding between ß3 in the N-terminal domain and ß4 in the C-terminal domain. This separation of domains is also manifested by the distribution of the temperature factors in each domain). The residues in the N-terminal domain have higher temperature factors (an average of $42Å^2$), and thus are more mobile than those in the C-terminal domain (an average of $33Å^2$). This high mobility of the N-terminal domain may have a functional significance in that it is in this domain where the catalytic site of GTP hydrolysis, the putative effector region (6) and the GTPase activating protein binding region (residues 30-40) (7,8) are located.

The differences between the two structures are mainly localized in the loops L1, L2, L3, and L4 in the N-terminal half of the molecule. The root-mean-square difference in Cα between p21(Val-12) and the normal protein is 1.26Å for residues in the N-terminal domain (residues 1-75) and 0.56Å for residues in the the C-terminal domain (residues 76-171). At the current stage of the refinement, the two regions with the largest differences are located in loops L1 and L4. However, the electron density for residues in loop L4 is not as well defined (evidenced by their high temperature factors). Further refinement is required to assess the significance of the differences in this region.

The most clear and largest structural differences were found in loop L1, corresponding to residues 9 through 18, which wraps around the ß-phosphate of the bound GDP molecule (Fig. 3). The current refinement results of the normal protein as well as the transforming p21(Val-12) reveal that there are several unusual aspects about this loop. First, there appear to be no side chains, except that of Lys-16, involved in binding to the phosphate. Second, the backbone amide groups are pointing toward the ß-phosphate, making hydrogen bonds to the phosphate oxygens and thus fixing the orientation of the ß-phosphate, which is presumably important for GTPase activity. Third, there is a metal ion (probably Mg^{++} or Ca^{++}) coordinated to the oxygen atoms of the ß-phosphate.

The simplest description of the structural differences in loop L1 is that the size of this loop in p21(Val-12) is much larger than that in the normal protein (Fig. 3), resulting in the loss of two hydrogen bonds (from the backbone NH groups of residues 12 and 13) to the ß-phosphate. We have suggested (2) that this loop would have straddled the phosphodiester bond between the ß- and γ-phosphates of GTP, and therefore is the prime candidate to be the catalytic site for GTP hydrolysis in the normal p21 protein. The loss of two hydrogen bonds may alter the orientation of the ß-phosphate and γ-phosphate when presented to an attacking group, thus changing the GTP hydrolysis rate. The position of the bound metal ion is such that it is a reasonable candidate for providing the attacking group (probably one of its bound water molecules) to the ß-phosphate, perhaps for the on-line displacement of the γ-phosphate, or the cation may be participating in fixing the ß-phosphate in collaboration with backbone NH groups, with an as yet unidentified water molecule attacking the ß-phosphate. An alternative mechanism, similar to that proposed for elongation factor Tu, is that the γ-phosphate (which is not present in our structure) is attacked, possibly by a water molecule

In summary, the crystal structure of transforming c-H-*ras* p21(Val-12, 1-171) with a bound GDP molecule provides a plausible explanation for the loss of GTPase activity of the protein, thus keeping the protein in the GTP bound state, which is thought to signal continued cell proliferation without regulation. A complete understanding will have to await the structure determination of the GTP bound c-H-*ras* p21 and of different activated mutants. Efforts to obtain p21 crystals with bound non-hydrolizable GTP analog and to determine the structure of the activated mutant p21(Leu-61, 1-171) are in progress.

ACKNOWLEDGEMENT

This work was supported also by a gift from Merck, Sharp and Dohme to the University of California-Berkeley. Sh.N. was supported by a fellowship from the Foundation for Promotion of Cancer Research.

REFERENCES

1. Barbacid, M (1987). *Ann Rev Biochem* **56**:779.
2. De Vos, AM (1988) *et al. Science* **239**:888.
3. Gibbs JB, Sigal IS, Poe M & Scolnick EM (1984). *Proc Nat Acad Sci USA* **81**:5704.
4. Miura, K *et al* (1986). *Jpn J Cancer Res (Gann)* **77**:45.
5. Jancarik, J *et al* (1988) *J Mol Biol* **200**:205.
6. Sigal IS, Gibbs JB, D'Alonzo JS & Scolnick E (1986). *Proc Nat Acad Sci USA* **83**:4725.
7. Calés C, Hancock J., Marshall C. & Hall A. (1988). *Nature* **332**:548.
8. Adari M, Lowy DR, Willumsen BM, Der CJ & McCormick F (1988). *Science* **240**:518.

Frontiers of NMR in Molecular Biology, pages 155-166

NMR STUDIES OF LYSOZYME: ANALYSIS OF COSY SPECTRA USING SIMULATIONS

Christina Redfield

Inorganic Chemistry Laboratory, University of Oxford
Oxford, OX1 3QR, England

INTRODUCTION

The virtually-complete sequential assignments available for hen and human lysozymes provide the starting point for a variety of studies (1). These include the measurement and comparison of spectral parameters including chemical shifts, coupling constants and linewidths for the two proteins, and the determination of the three-dimensional structure of lysozyme in solution using NOE distance restraints and coupling-constant torsion-angle restraints.

A great deal of information can be obtained from the fingerprint region of a COSY spectrum of lysozyme. This part of the spectrum contains one cross peak for each amino acid residue in the protein. Thus, it is possible to obtain information for each residue or to monitor the effect of a perturbation throughout the protein simply by analyzing the fingerprint region of the COSY spectrum. For example, information about hydrogen bonding and protein dynamics can be obtained by measuring hydrogen-exchange rates or by determining the temperature coefficients of amide chemical shifts. Backbone torsion-angle restraints, important in protein structure calculations, are obtained from the NH-αCH coupling constants extracted from the anti-phase components of fingerprint region COSY peaks. Information about the electrostatic environment of residues in the protein can be obtained by following the change in chemical shift with pH.

The experimental parameters used to collect a COSY data set will depend on the type of information to be extracted from the spectrum. If hydrogen-exchange rates are to be measured in real time from cross-peak intensities then the most important factor may be to collect the 2-D data set as rapidly as possible at the expense of digital resolution and signal-to-noise. If, on the other hand, *NH*-α*CH* coupling constants are being measured then digital resolution and a good signal-to-noise ratio will be far more important than rapid data collection. In this paper factors which influence the choice of spectral parameters will be discussed. Spectral simulations are used to assess the importance of various experimental parameters and to extract spectral parameters from COSY spectra.

RESULTS AND DISCUSSION

The full phase-cycling scheme for a COSY experiment consists of 8 transients per t_1 increment (2). Sign discrimination on the F1 domain is achieved by repeating these 8 transients with the first pulse phase-shifted by 90° giving a total of 16 transients per t_1 increment (3,4). With a recycle delay of 1 second and a $t_{1_{max}}$ of 73 ms an experiment can usually be collected in ~3 hours. 2-D spectra presented in the literature are often collected with additional signal averaging and experiments are recorded for 12 hours or longer. Many researchers present DQF COSY spectra rather than COSY spectra. The minimum phase-cycling in a DQF COSY requires 32 transients per t_1 increment and the signal-to-noise ratio is a factor of 2 worse than for a COSY spectrum (5). The major advantage of DQF COSY, the improved diagonal-peak shape, is not important when analyzing the fingerprint region because this part of the spectrum is well separated from the diagonal. In order to decrease the acquisition time for a COSY spectrum the 4-step phase-cycling for the removal of quadrature images is eliminated (6). Removal of this phase-cycling leads to an unattractive 'anti-diagonal' artefact in the 2-D spectrum. However, this artefact, perpendicular to the diagonal, covers regions of the spectrum which do not ordinarily contain cross peaks and, therefore, does not obscure information. If the relaxation delay is shortened to 0.75 s then a COSY spectrum with the remaining 2-step phase cycling for axial-peak suppression and $t_{1_{max}}$ of 73 ms can be collected in 50 minutes. If only half the number of t_1 increments is collected ($t_{1_{max}}$=36 ms) the acquisition time is shortened to 25 minutes.

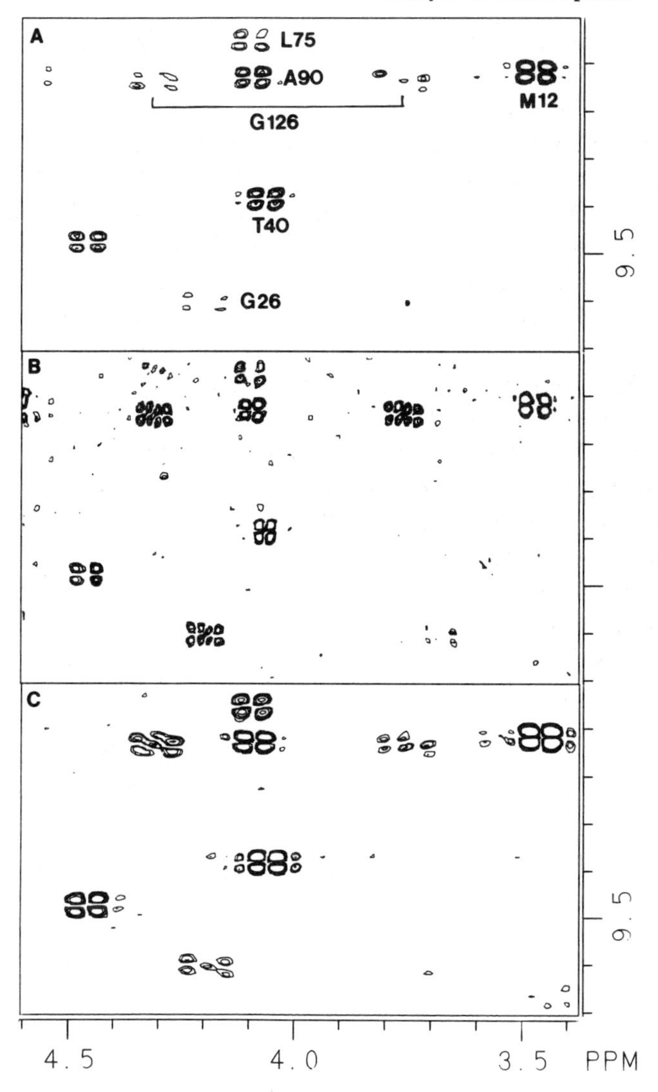

FIGURE 1. Part of the 500 MHz phase-sensitive COSY spectrum of 7mM hen egg-white lysozyme in 90% H_2O/10% D_2O at pH 3.8. The spectra in A), B) and C) were collected respectively: in 25 minutes with $t_{1_{max}}$=36 ms and 4 transients per t_1 increment; in 50 minutes with $t_{1_{max}}$=73 ms and 4 transients per t_1 increment; and in 42 minutes with $t_{1_{max}}$=36 ms and 8 transients per t_1 increment. All three data sets were zerofilled to give a final digital resolution of 3.44 Hz/point in both the F1 and F2 dimensions.

A part of the fingerprint region of a COSY spectrum of 7 mM lysozyme obtained in 25 minutes is shown in Figure 1A. This spectrum contains cross peaks for about 90 % of the residues of lysozyme. If the total acquisition time is doubled then either twice as many t_1 increments or twice as many transients can be collected. A comparison of Figures 1A-C highlights the effects of altering these two parameters. Doubling the number of t_1 increments leads to an improvement in the appearance and information content of the cross peaks. However, because t_1 increments in which the signal has decayed considerably are collected the signal-to-noise ratio in the spectrum has decreased. It should be stressed that the data in Figures 1A-C have been zero filled to give the same final digital resolution in both dimensions as summarized in the caption. The most startling difference between Figures 1A and B is the appearance of the glycine cross peaks. In Figure 1B the cross peaks of G26 and G126 are relatively intense and have the 8-component pattern characteristic of coupling of the NH to two αCH resonances. In Figure 1A this characteristic pattern is absent and the cross peaks are of low intensity. Another clear feature of the spectra is the shape of cross peaks from residues other than glycine. In Figures 1A and 1C all the peaks have the same rectangular appearance whereas in Figure 1B some cross peaks, like T40 and A90, have a square shape while others, like M12 and L75, are rectangular. Spectral simulations can be used to explain the cross-peak patterns and shapes observed.

Cross-peak simulations for G26, T40 and M12 using $t_{1_{max}}$ values of 36 ms and 73 ms are compared with experimental peaks in Figures 2A-C. Good agreement between simulated and experimental peaks is achieved for all three residues. When a $t_{1_{max}}$ value of 36 ms is used the four central components of the glycine cross peak cancel leaving only the four outer components and a reduced overall cross-peak intensity. Longer values of $t_{1_{max}}$ are required to resolve the eight anti-phase components characteristic of the glycine cross peak. Therefore, if NMR parameters for glycine residues are of interest in a study it will be important to collect data with longer $t_{1_{max}}$ values at the expense of signal-to-noise. If glycine residues are not of interest then better signal-to-noise can be achieved with shorter values of $t_{1_{max}}$.

Differences in the appearance of non-glycine cross peaks also result from differences in the resolution in the t_1 domain. All cross peaks have the same rectangular shape regardless of the value of the passive αCH-βCH coupling constant when a $t_{1_{max}}$ value of only 36 ms is used. The square pattern observed for T40, when a $t_{1_{max}}$ value of 73 ms is used, reflects the small passive α-β coupling constant of ~4 Hz. Simulations for a range of NH-αCH and αCH-βCH coupling constants, and a $t_{1_{max}}$ value of 73 ms, show that the square pattern is always obtained when the single αCH-βCH coupling constant or both αCH-βCH coupling constants are small. Cross peaks in spectra collected with a $t_{1_{max}}$ value of 73 ms have a rectangular shape if at least one αCH-βCH coupling constant is large. Thus the shape of the NH-αCH cross peak can give some information about the size of the passive αCH-βCH couplings and, therefore, can provide useful information when making spin-system identifications and resonance

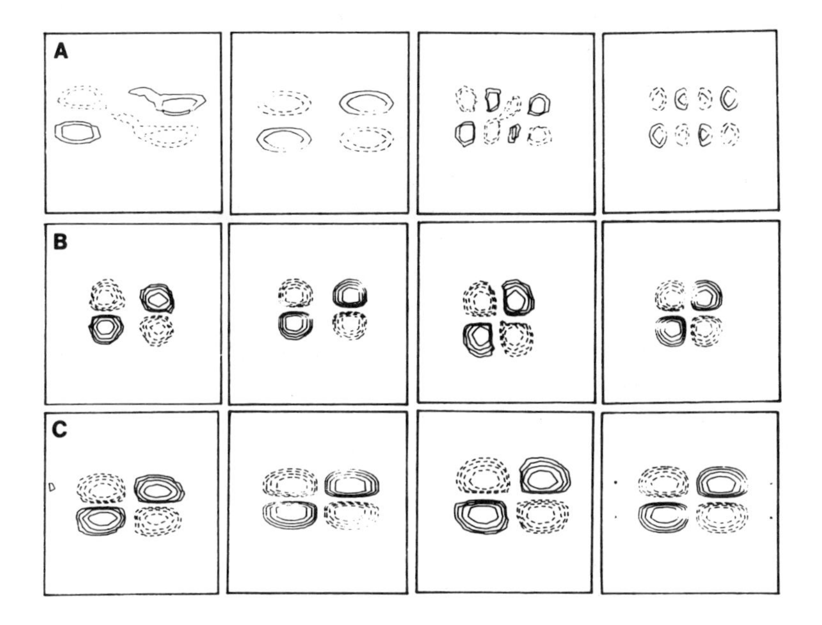

FIGURE 2. A comparison of experimental and simulated cross peaks for residues A) G26, B) T40 and C) M12. The spectra in the two left-hand columns correspond to $t_{1_{max}}$=36 ms and those in the two right-hand columns correspond to $t_{1_{max}}$=73 ms. The spectra in the first and third columns are experimental whilst those in the second and fourth columns are simulated. The horizontal and vertical dimensions represent F1 and F2 respectively. All spectra were zerofilled to give a final digital resolution of 3.44 Hz/point in both dimensions. The coupling constants used in the simulations were: A) $J_{N\alpha_1}$ = 6.0, $J_{N\alpha_2}$ = 7.0, $J_{\alpha_1\alpha_2}$ = 18.0 Hz; B) $J_{N\alpha}$ = 4.1, $J_{\alpha\beta}$ = 3.8 Hz; and C) $J_{N\alpha}$ = 4.8, $J_{\alpha\beta}$ = 12.0, 3.2 Hz. Cross-peak patterns were calculated using the program SIMULATION.

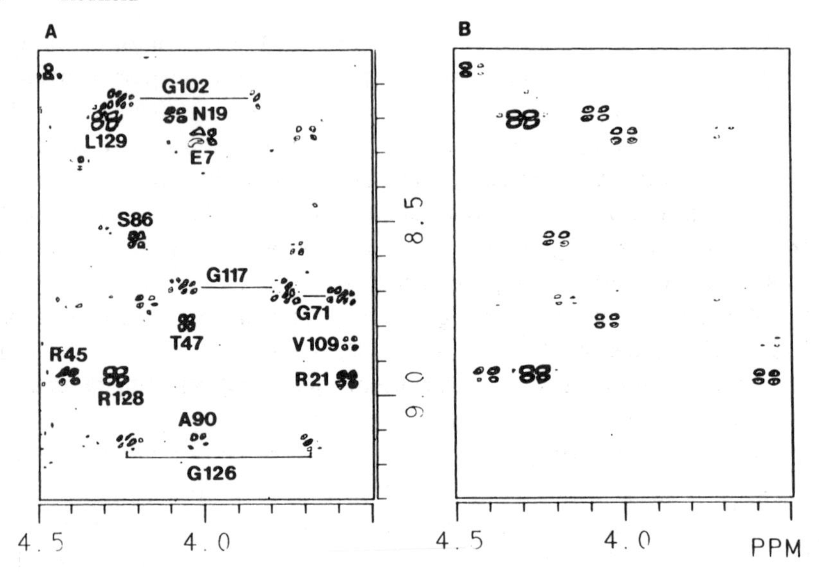

FIGURE 3. Part of the 500 MHz phase-sensitive COSY spectrum of 7mM hen lysozyme. Freeze-dried fully-exchanged lysozyme was dissolved in 90% H_2O/10% D_2O at pH 2.5 just prior to data collection. The spectra in A) and B) were collected using 4 transients per t_1 increment and with $t_{1_{max}}$ values of 73 ms and 36 ms respectively. Total acquisition times for A) and B) were 50 mins and 25 mins respectively. Cross peaks which are labelled in the spectrum correspond to the most rapidly exchanging amides at pH 2.5.

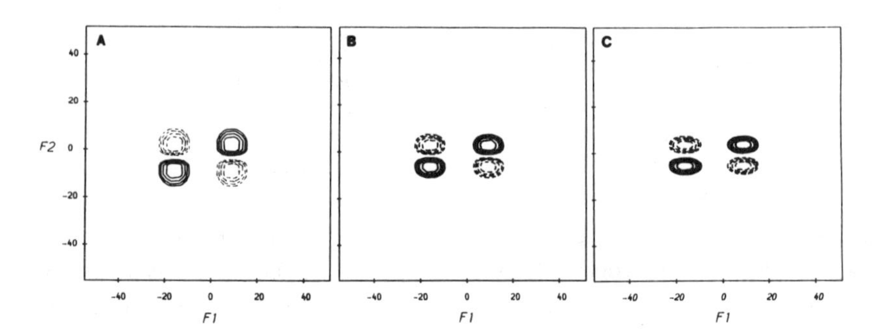

FIGURE 4. Simulations of the fingerprint-region cross peak of D52 for different values of $t_{2_{max}}$. All spectra are zerofilled to give digital resolutions of 3.44 Hz/point in F1 and A) 3.44, B) 1.72 and C) 0.86 Hz/point in F2. Spectra are simulated with $t_{1_{max}}$ = 73 ms and A) $t_{2_{max}}$ = 145 ms, B) $t_{2_{max}}$ = 291 ms, and C) $t_{2_{max}}$ = 582 ms.

assignments. It can also be useful when following changes in cross-peak positions in crowded regions of the spectrum as a function of, for example, temperature or pH. In these studies spectra should be collected with a $t_{1_{max}}$ value of 73 ms. However, collection of a large number of t_1 values would not be necessary if NMR measurements were being made for previously-assigned non-glycine residues which were not shifting in the spectrum during the course of the series of experiments.

Application of fast COSY techniques to the study of hydrogen exchange in lysozyme is illustrated in Figure 3. Freeze-dried lysozyme in which all labile hydrogens had been exchanged with deuterium was dissolved in 90% H_2O/10% D_2O, and a COSY data set collected immediately (1). A part of the fingerprint region of the COSY spectrum collected in 50 min with $t_{1_{max}}$ of 73 ms is shown in Figure 3A. Cross peaks visible in the spectrum correspond to the most rapidly exchanging amides at pH 2.5 and 25°C. It is clear from Figure 3A that several glycines fall into this hydrogen-exchange category. These cross peaks are not visible in the spectrum collected with $t_{1_{max}}$ of 36 ms shown in Figure 3B, and interpretation of this spectrum would lead to the incorrect conclusion that glycine residues are not in the group of most rapidly exchanging amides.

The simulations and spectra presented thus far have dealt with the question of resolution in the t_1 domain. Now consider the problem of resolution in the t_2 domain. Simulations of the D52 cross peak for various values of $t_{2_{max}}$ are shown in Figure 4. In contrast to the F1 dimension where both active (NH-αCH) and passive (αCH-αCH, αCH-βCH) couplings affect the appearance of the peak; the F2 dimension is influenced only by the active NH-αCH coupling and cross sections parallel to F2 are anti-phase doublets. Inspection of Figure 4 shows that the appearance of the cross peak does not change dramatically as the value of $t_{2_{max}}$ is increased. The only noticeable difference is the improved resolution of the positive and negative cross peak components. Signal-to-noise decreases in the 2-D spectrum as $t_{2_{max}}$ is increased. Therefore, only when resolution in F2 is critical, such as in the measurement of NH-αCH coupling constants, should long $t_{2_{max}}$ values be used. Because good signal-to-noise will be important in these measurements sufficient signal averaging is essential and data sets cannot usually be collected rapidly.

NH-αCH coupling constants can be measured from cross sections parallel to F2 through fingerprint-region peaks (4). These cross sections contain anti-phase doublets in which the observed splitting between positive and negative components provides a measure of the coupling constant, J, as shown in Figure 5. Because of the anti-phase nature of the cross peak the observed splitting will be greater than the true coupling constant (7). This discrepancy increases as the linewidth increases and is particularly bad for small coupling constants as shown in Figure 6 (7). For small proteins such as BPTI where linewidths are generally smaller than J, coupling constants can be measured directly from the observed splittings (4). However, for somewhat larger proteins such as lysozyme where amide linewidths are on the order of 6-10 Hz, corrections must be made for the effects of anti-phase cancellation. It can be seen from Figure 6 that several pairs of linewidth and coupling constant can give rise to the same observed splitting. For example, the observed splitting of 5.7 Hz for V99 in Figure 5 could arise from a true J of 5.5 Hz and a linewidth of 6 Hz, a true J of 5.3 Hz and a linewidth of 7 Hz and so on. One approach would be to assume an average linewidth for all amides in the protein and to use the appropriate curve in Figure 6 to make corrections. However, analysis of the large range of peak heights observed in the 2-D spectra of lysozyme indicates that the protein exhibits a range of linewidths (1). A more satisfactory approach is to use spectral simualtions to fit the observed peak shape.

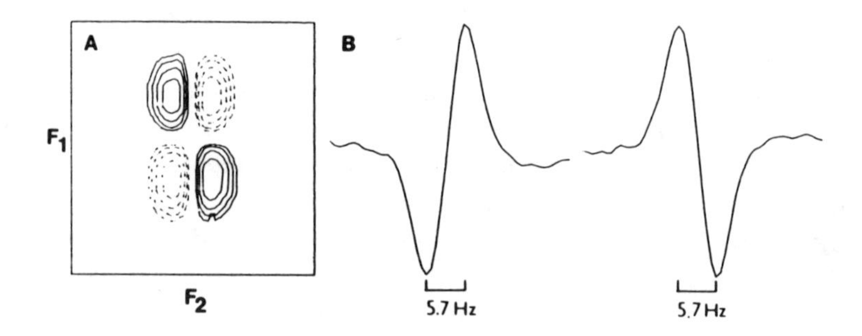

FIGURE 5. Fingerprint-region A) cross peak and B) cross sections parallel to F2 for V99. The COSY spectrum was collected at 55°C for a 5 mM sample of hen lysozyme at pH 3.8. The values of $t_{1_{max}}$ and $t_{2_{max}}$ were 91 ms and 729ms respectively. The data were zerofilled to give digital resolutions of 2.74 Hz/point in F1 and 0.68 Hz/point in F2. The spectrum was collected in 18 hours with 64 scans per t_1 increment.

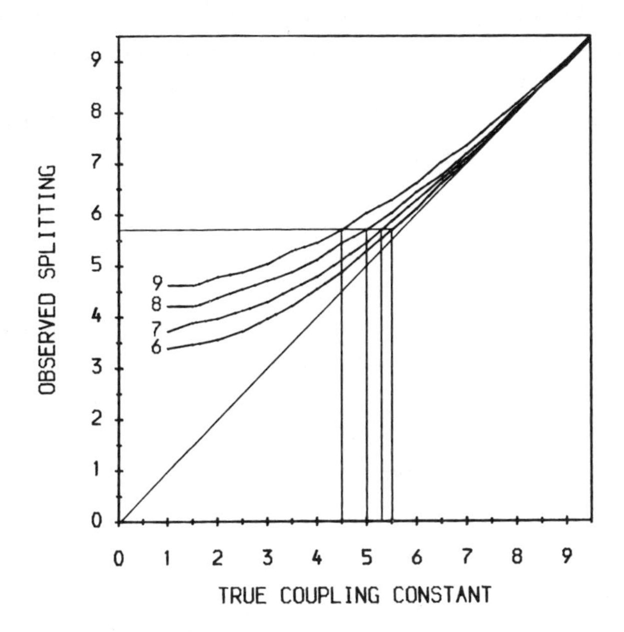

FIGURE 6. A plot of observed splittings versus true coupling constants for linewidths of 6-9 Hz. The observed splitting of 5.7 Hz for V99 would correspond to true coupling constants of 5.5, 5.3, 5.0 and 4.5 Hz at linewidths of 6, 7, 8 and 9 Hz respectively.

Cross sections are simulated using pairs of linewidth and coupling constant and the experimental digital resolution, resolution enhancement and zerofilling. These simulations are then compared with the experimental cross sections. The values of linewidth and coupling constant are changed and the simulation repeated until a good agreement is achieved. The results of such a fit are shown in Figure 7 for a cross section of V99. The best fit, shown in Figure 7B, is achieved with a NH-αCH coupling constant of 5.2 Hz and a linewidth of 7.0 Hz. The difference between the experimental cross section and the simulation is compared with cross-section noise in Figure 7C. The rms difference of 0.064 is close to the observed rms noise of 0.062. If the simulation is carried out with a coupling constant of 5.4 Hz and a linewidth of 6.0 Hz, significantly worse agreement is found (rms = 0.114) as shown in Figure 7D. Using this procedure corrected coupling constants and linewidth values have been obtained for both hen and human lysozymes. The linewidth values found for the amides of lysozyme range from 4.5 to 12.5 Hz at 55°C. The coupling constants are currently being used as torsion-angle restraints in calculations of the structure of lysozyme in solution.

The examples presented in this paper show that the parameters used for collecting a COSY spectrum should be chosen by considering

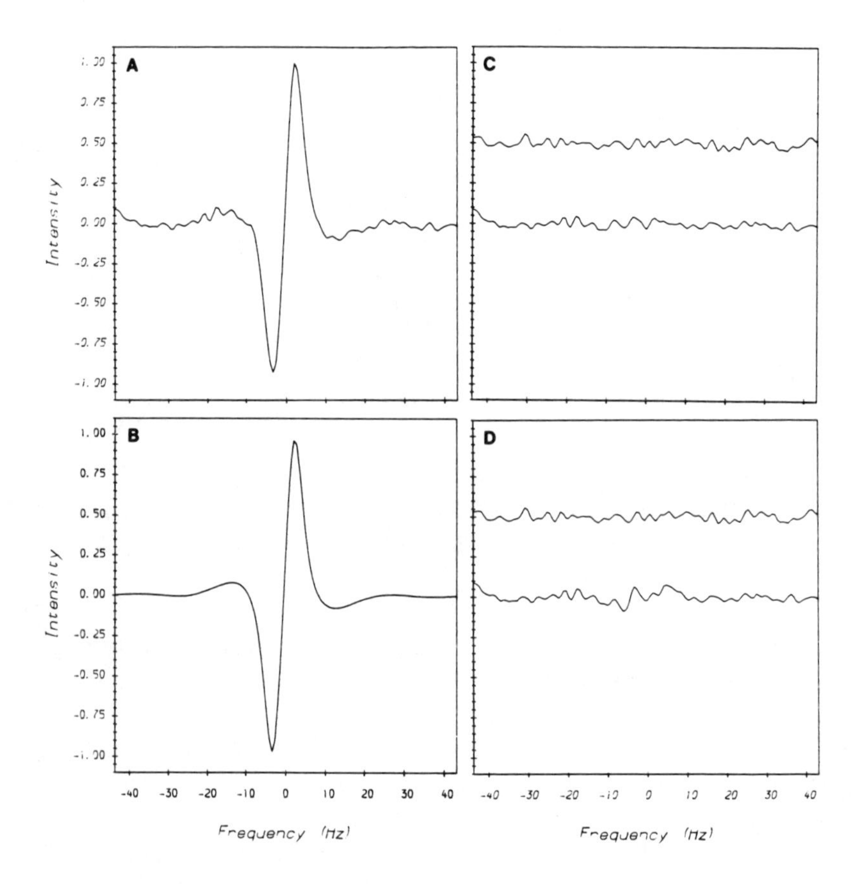

FIGURE 7. Comparison of experimental and simulated F2 cross sections for V99. A) Experimental cross section. B) Best-fit simulation obtained with J=5.2 Hz, LW=7.0 Hz. C) Comparison of the difference between experimental and best-fit cross sections (lower trace) with the cross-section noise level (upper trace). D) Comparison of the difference between the experimental cross section and a simulation obtained with J=5.4 Hz and LW=6.0 Hz (lower trace) with the cross-section noise level (upper trace).

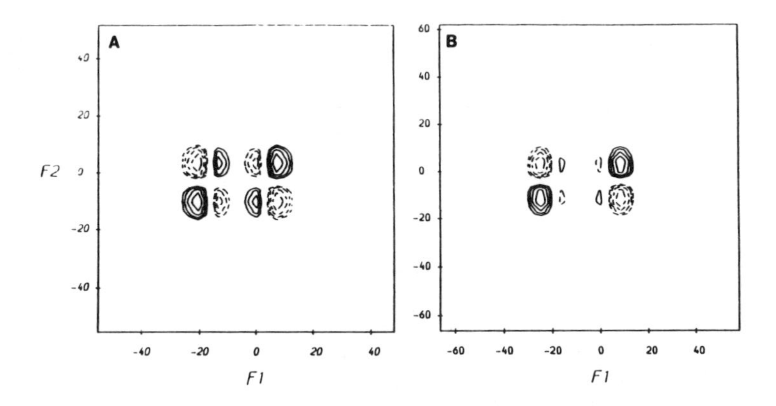

FIGURE 8. Simulated glycine *NH-αCH* cross peaks at A) 500 MHz and B) 600 MHz. Simulations were carried out with sweep widths of A) 7042.25 Hz and B) 8474.6 Hz in both dimensions. Both cross peaks were simulated with 512 real points in the t_1 domain and 1024 real points in the t_2 domain. After zerofilling, digital resolution in F1 and F2 were respectively: A) 6.88 and 3.44 Hz/point and B) 8.28 and 4.14 Hz/point.

the type of information which is required from the spectrum. Use of spectral simulations can be important in predicting cross-peak patterns and in choosing appropriate parameters (8). In this paper I have concentrated on the fingerprint region of the spectrum, a spectral region rich in information. Similar simulations can be used to choose the best parameters for collecting and processing other regions of COSY spectra or of other spectra such as NOESY or HOHAHA. Simulations can also be used to assess the effects of increases in sweep width which are required when using the higher-field spectrometers now becoming available. Consider the simulations shown in Figure 8; a glycine cross peak is simulated at 500 MHz and 600 MHz using identical data matrix sizes. The effect of a 20% increase in the sweep width on resolution in F1 is clear. The glycine cross peak simulated at 600 MHz has lost its characteristic 8-component pattern. Using spectral simulation it can be shown that when more t_1 increments are collected at 600 MHz the desired cross-peak pattern is recovered.

ACKNOWLEDGEMENTS

This contribution is from the Oxford Centre for Molecular Sciences which is funded by the Science and Engineering Research Council of Great Britain.

REFERENCES

1. Redfield C, Dobson C M, (1988). Sequential ^{1}H NMR Assignments and Secondary Structure of Hen Egg White Lysozyme in Solution. Biochemistry 27:22.
2. Bax A, Freeman F, (1981). Investigation of Complex Networks of Spin-Spin Coupling by Two-Dimensional NMR. J. Magn. Reson. 44:542.
3. States D J, Haberkorn R A, Ruben D J, (1982). A Two-Dimensional Nuclear Overhauser Experiment with Pure Absorption Phase is Four Quadrants. J. Magn. Reson. 48:286.
4. Marion D, Wüthrich K, (1983). Application of Phase Sensitive Two-Dimensional Correlated Spectroscopy (COSY) for Measurements of ^{1}H-^{1}H Spin-Spin Coupling Constants in Proteins. Biochem. Biophys. Res. Commun. 113:967.
5. Piantini U, Sørensen O W, and Ernst R R, (1982). Multiple Quantum Filters for Elucidating NMR Coupling Networks. J. Am. Chem. Soc. 104:6800.
6. Hoult D I, Richards R E, (1975). Critical Factors in the Design of Sensitive High Resolution Nuclear Magnetic Resonance Spectrometers. Proc. Roy. Soc. London Ser. A. 344:311.
7. Neuhaus D, Wagner G, Vašák M, Kägi J H R, Wüthrich K, (1985). Systematic Application of High-Resolution, Phase-sensitive Two-Dimensional ^{1}H NMR Techniques for the Identification of the Amino-acid Proton Spin Systems in Proteins. Eur. J. Biochem. 151:257.
8. Widmer H, Wüthrich K, (1986). Simulation of Two-Dimensional NMR Experiments using Numerical Density Matrix Calculations. J Magn Reson 70:270.

Frontiers of NMR in Molecular
Biology, pages 167-175

COMPUTER ASSISTED ASSIGNMENT OF TWO-DIMENSIONAL NMR SPECTRA OF PROTEINS

Jeffrey C. Hoch*, Michael M. Burns*, and Christina Redfield†

*Rowland Institute for Science
100 Cambridge Parkway
Cambridge, MA 02142

†Inorganic Chemistry Laboratory
University of Oxford
Oxford, England OX1 3QR

There is little disagreement that much of the analysis of NMR data from experiments on proteins is routine and tedious. The computation of the spectrum from the free induction decay, the identification of spectral features and the estimation of parameters characterizing those features, and the correlation of spectral parameters to assign the features to specific parts of the molecule are all straightforward, at least in principle. In practice the complexity of proteins makes the analysis of their NMR spectra a tedious task (through the sheer number of peaks present) and difficult (due to overlap). It is not suprising, then, that automated analysis of two-dimensional spectra has become an often-stated goal.

While significant progress toward that goal has been made over the past few years, it is important to recognize that a less ambitious goal, that of *computer-assisted* analysis of the NMR spectra of proteins, nevertheless represents a significant step, since the assignment of the spectrum is often the rate-limiting step in the determination of protein structures by NMR. Toward that goal, we have implemented a "clerk system" for aiding the assignment process. A clerk system is one which allows an expert—in this case the spectroscopist—to use his or her judgement, but frees the spectroscopist from worrying about tedious details.

Our system consists of a set of software tools which fall into two categories. One set is implemented within a program we call LightBox, which requires a display list graphics output device (e.g. Evans & Sutherland or Megatek). LightBox provides tools for displaying (including registration or overlaying) several spectra, for locating peaks in the spectra, and for classifying spin systems by finding correlations between spectra (or sub-spectra). The second set of tools is implemented in a LISP environment. These tools use the list of spin systems produced by LightBox to generate possible assignments which are consistent with the primary sequence of the protein.

The development process was guided by the experience gained in assigning the ^1H spectrum of lysozyme from hen egg-white[1]. Most of the tools, however, are sufficiently general to aid other sequential assignment strategies.

The major components of LightBox provide capabilities for:

Display. Several 2D contour plots can be displayed simultaneously, with the ability to interactively scale and translate spectra relative to one another. Interactive cursors can be attached to each of the displayed contour plots; cursors can be "bound" to one another (for example, the f_2 coordinate of one cursor can be bound to the f_1 coordinate of another cursor, so that moving the second cursor in f_1 also moves the first cursor in f_2). Bindings are designated by the colors used to draw the cursors. A "dump" facility provides a means of generating color hardcopy of the screen on a pen plotter.

Feature Locators. Implemented within LightBox are simple peak-picking algorithms and procedures for using symmetry recognition[2] to locate antiphase COSY multiplets. Symmetry recognition is sufficiently computationally intensive that it is not interactive on a VAX 8250, however. While improvements in computational speed (such as the new class of "superworkstations") will make interactive symmetry recognition feasible, in the mean time we have implemented a less sophisticated method for locating antiphase multiplets, consisting of the following steps:

(1) Locate all peaks with magnitude larger than *threshold*.

(2) For each point falling within $(\pm\Delta\omega_1, \pm\Delta\omega_2)$ of a peak located in (1), store the *sign* of the peak and the *quadrant* of the point relative to the peak (e.g. upper right = northeast, etc.).

(3) Create a bit image (each pixel either 1 or 0) in which only those points in (2) which correspond to −NE, −SW, +NW, and +SE are set to one; all other pixels are set to zero.

(4) Compute the center of mass for all contiguous regions[3] in the bit image generated in (3). These are (approximately) the centers of the antiphase multiplets.

Spin System Identification. Once peak lists are available, it is a straightforward matter to search the lists for classes of spin systems. Using COSY, RELAY, and double RELAY data the spin systems can be identified and assigned to one of 10 classes. As an example, a LightBox session for identifying alanine spin systems in lysozyme from hen egg-white is depicted in Figures 1–3. COSY data for the α-methyl region is loaded into the first window, contours are drawn (±5.0) and peaks above |5.0| are located. The second window is loaded with the fingerprint region of the COSY spectrum, contours are drawn (±1.0), and peaks above |1.0| are located. The methyl-H^N region of the RELAY spectrum is loaded into the third window, contours drawn (±1.0), and peaks above |1.0| are located. The state of the graphics display following these steps is shown in Figure 1 (for this publication black and white depictions are shown). Next the antiphase multiplets present in each window are located using the procedure described above (Figure 2). Finally, LightBox searches for alanine spin systems by identifying resonances for which crosspeaks consistent with an amide-α-(β-methyl) spin system are found in the appropriate locations (Figure 3), using a tolerence of 0.025 ppm (chemical shifts within 0.025 ppm of one another are assumed to be the same). The lists generated by LightBox, whether of peaks, multiplets, or spin systems, can be manually edited by the spectroscopist.

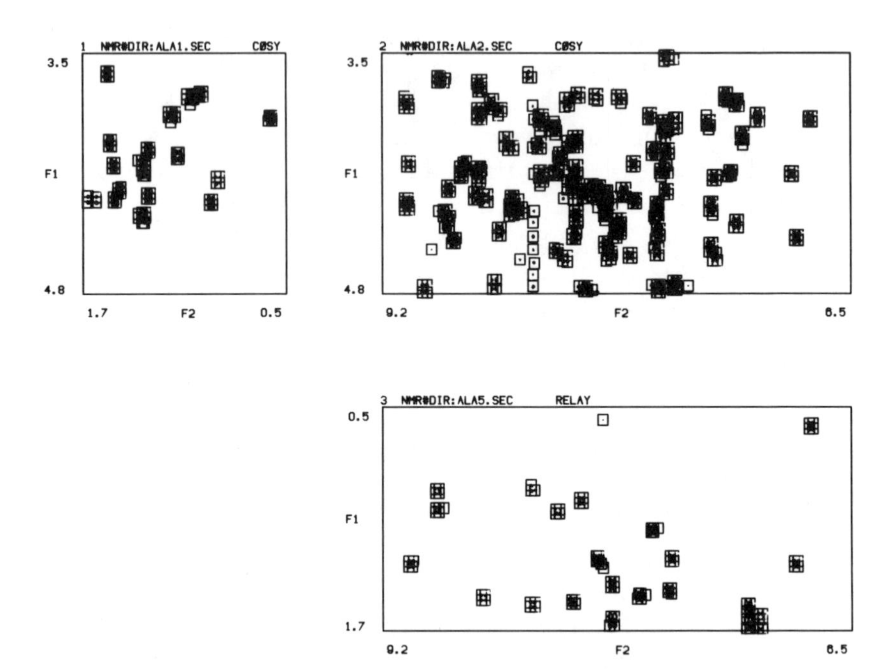

Figure 1.

The LISP environment for performing sequential assignment employs simple enumeration and pattern matching algorithms. Input data consists of the chemical shifts of COSY fingerprint region crosspeaks (identified by the class of the amino acid), the chemical shifts of NOESY H^{α}-H^{N} and H^{N}-H^{N} cross peaks, and the protein sequence. The search for possible assignments consistent with the protein sequence is based on lists of all possible linkings of COSY fingerprint region cross peaks to NOESY cross peaks within a given tolerance and of all possible pairings of COSY fingerprint cross peaks via the NOESY links.

To keep the number of possibilities manageable, the search is constrained using parameters such as the minimum length of the chain, the minimum number of "low frequency" classes (containing a single amino acid) in the chain, and whether H^{α}-H^{N}, H^{N}-H^{N}, or both types of NOESY links can be used to forge the chain.

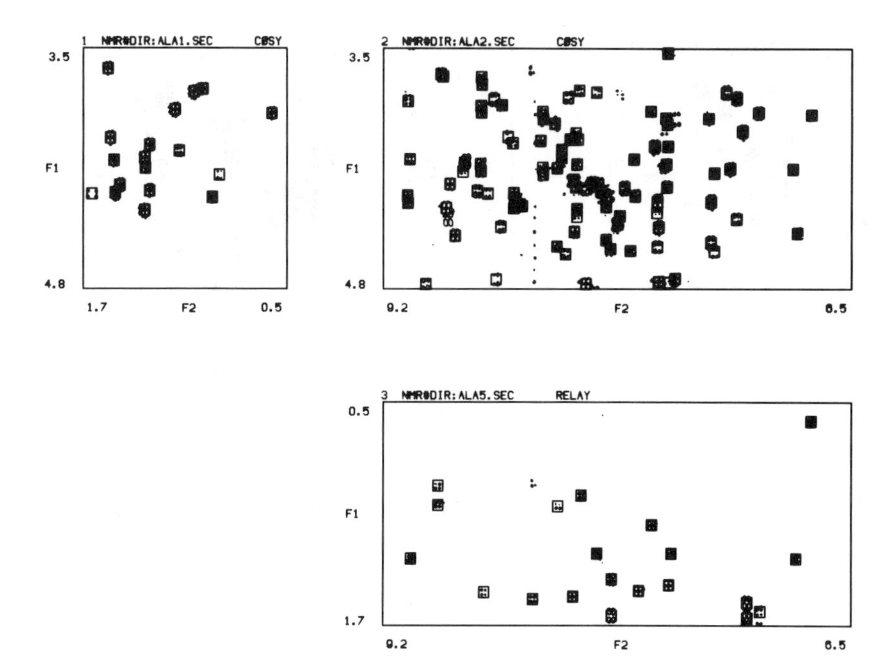

Figure 2.

The longer the chain and the higher the number of low frequency classes, the higher the likelihood that the chain is unique. For lysozyme, where the longest NOESY-linked chain is 12 residues long, a sequential assignment strategy can be implemented by assigning the long chains first, then proceeding to assign successively shorter chains containing fewer low frequency classes. The following script, based on data for lysozyme, illustrates how the tools can be used to implement such a strategy. The number in parentheses is the residue number of the first peak in the chain. The remaining numbers are the peak numbers of the COSY fingerprint cross peaks, numbered here by increasing residue number to illustrate the types of ambiguities that arise.

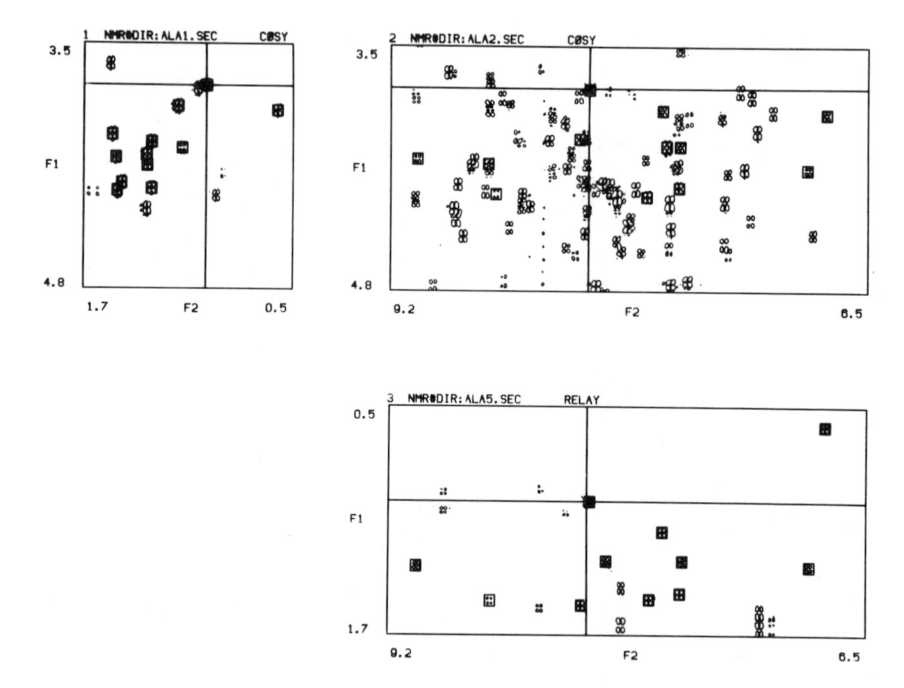

Figure 3.

(all-seqs :thru any :length 10 :lofreqs 3)

Summary of chains found:

(6) 5 6 7 8 9 10 11 12 13 14

(6) 5 6 7 8 9 10 11 12 13 107

(39) 40 41 42 43 44 45 46 47 48 49

(90) 89 90 91 92 93 94 31 96 97 98 99 100

(90) 89 90 91 92 93 94 95 96 97 98 99 100

With the exception of residues 15 and 96, these chains yield unambiguous assignments. Assigning only the unambiguous COSY and linking NOESY cross peaks (not listed), the search reveals:

(all-seqs :thru any :length 7 :lofreqs 3)

Summary of chains found:

(16) 15 17 18 19 20 21 22

(16) 15 17 18 19 20 21 22

(16) 15 17 18 19 20 115 116

(49) 50 52 53 54 55 56 58

(115) 20 115 116 118 119 120 121 122

Here a new type of ambiguity arises: we could either have the chain 20 21 22 or 20 115 116. Whenever ambiguities such as this arise, the state of all variables in the LISP environment can be saved, allowing branching searches. Since the chain 20 115 116 "fits" two positions in the sequence, whereas 20 21 22 fits only one (it appears twice because 21 and 22 are linked via both an H^{α}-H^{N} and an H^{N}-H^{N} NOESY cross peak), we choose the chain 20 21 22, and leave residue 115 unassigned (but assign 116 on). Proceeding with this branch, there are no longer any chains containing at least three low frequency classes. So we lower the low frequency requirement:

(all-seqs :thru any :length 7 :lofreqs 2)

Summary of chains found:

(31) 32 33 34 35 36 37 38 86 87

(31) 32 33 34 35 36 37 38 39

(31) 32 33 34 35 36 37 38 114

(31) 32 33 60 35 36 37 38 86 87

(31) 32 33 60 35 36 37 38 39

(31) 32 33 60 35 36 37 38 114

Here only residues 31, 32, and 34–37 have unique assigments. Continuing in this fashion, about two-thirds of the sequence can be assigned in a matter of a few hours. At that point, further progress demands more detailed reasoning, and more sophisticated tools will be required to assist this "end game".

There is a great deal of collective experience with sequential assignment strategies[4], and generalizations such as the importance of identifying residues as uniquely as possible have grown out of that experience. Just a few minutes with the assignment tools can make these generalizations concrete for a specific protein. For example, searching the lysozyme data reveals 17 chains at least 5 residues long having three low frequency classes. There are 54 chains at least 5 residues long with two low frequency classes. The average number of chains corresponding to a given portion of the protein sequence is 1.9 for the chains containing 3 low frequency classes and 3.7 for two low frequency classes; if all the chains were unique the average would be 1.

There are two main areas where further development is necessary to make clerk systems for assigning protein NMR spectra more useful, and expert systems—capable of performing at the level of a human expert—more feasible. One area is the development of more robust algorithms for feature detection and parameter estimation; humans are better at detecting overlapping resonances than algorithms currently in use. The other area concerns the ability of computer programs to utilize diverse types of information, such as coupling constants, pH and temperature dependence, and hydrogen exchange rates. The prospects are good that we will soon see significant progress in both of these areas. As the cost of computation declines, more sophisticated feature detection algorithms (including pattern and symmetry recognition) will become more feasible. Recent development of software tools for dealing with diverse forms of data (such as production system languages, object-oriented languages, and database-query languages) will provide the means for incorporating more knowledge about proteins into software systems for assigning protein NMR spectra.

REFERENCES

1. Redfield C and Dobson CM (*1988*). *Biochemistry* **27**:122-136.

2. Hoch JC, Hengyi S, Kjær M, Ludvigsen S, and Poulsen FM (*1987*). *Carlsberg Res Commun* **52**:111-122.

3. Winston PH and Horn BKP (*1984*). "LISP", 2nd Ed., Reading, Massachusetts: Addison-Wesley, Chapter 10.

4. Wüthrich K (*1986*). "NMR of Proteins and Nucleic acids," New York: Wiley-Interscience.

Frontiers of NMR in Molecular Biology, pages 177-187

DETERMINATION OF DNA AND PROTEIN STRUCTURES IN SOLUTION VIA COMPLETE RELAXATION MATRIX ANALYSIS OF 2D NOE SPECTRA[1]

Thomas L. James and Brandan A. Borgias

Departments of Pharmaceutical Chemistry and Radiology
University of California, San Francisco, CA 94143

ABSTRACT Two-dimensional nuclear Overhauser effect (2D NOE) spectral peak intensities are strongly dependent on interproton distances. Although the commonly used two-spin approximation leads to significant errors in distance determinations, a complete relaxation matrix analysis accounting for all dipolar interactions enables internuclear distances and structural features to be gained via iterative fitting of experimental 2D NOE spectra with theoretical spectra.

INTRODUCTION

Two-dimensional NMR is now routinely used in a vast array of specialized experiments for spectrum assignment and structural characterization of nucleic acids and proteins. Here we will principally examine the capability of using the homonuclear two-dimensional nuclear Overhauser effect (2D NOE) experiment; it has the potential for providing numerous interproton distances. The use of 2D NOE spectra in macromolecular structure determination is now becoming widespread. But, while considerable success has been achieved using this technique to provide distance constraints for distance geometry or molecular dynamics calculations leading to structures, there are certain limitations and precautions that are not always appreciated.

A proposed scheme which combines experimentally observable structural constraints with various calculational strategies to generate structures is shown in Figure 1. The scheme relies upon two recent developments applied in conjunction to provide a direct route to molecular structure in non-crystalline phases: 2D NMR and calculational strategies, e.g., the distance geometry (DG) algorithm (1,2), molecular mechanics (MM) (3), and molecular dynamics (MD)(3-5). The structure of any molecule can be determined with a sufficient number of structural constraints, e.g., internuclear distances and bond torsion angles, in conjunction with holonomic constraints of bond lengths, bond angles, and steric limitations. One can either include (MM, MD) or not include (DG) energetic considerations.

Problems addressed with DNA and with protein structure studies are often of a different nature. In general, we are interested in fairly subtle structural changes in the

[1] This work was supported by the National Institutes of Health via grant GM39247.

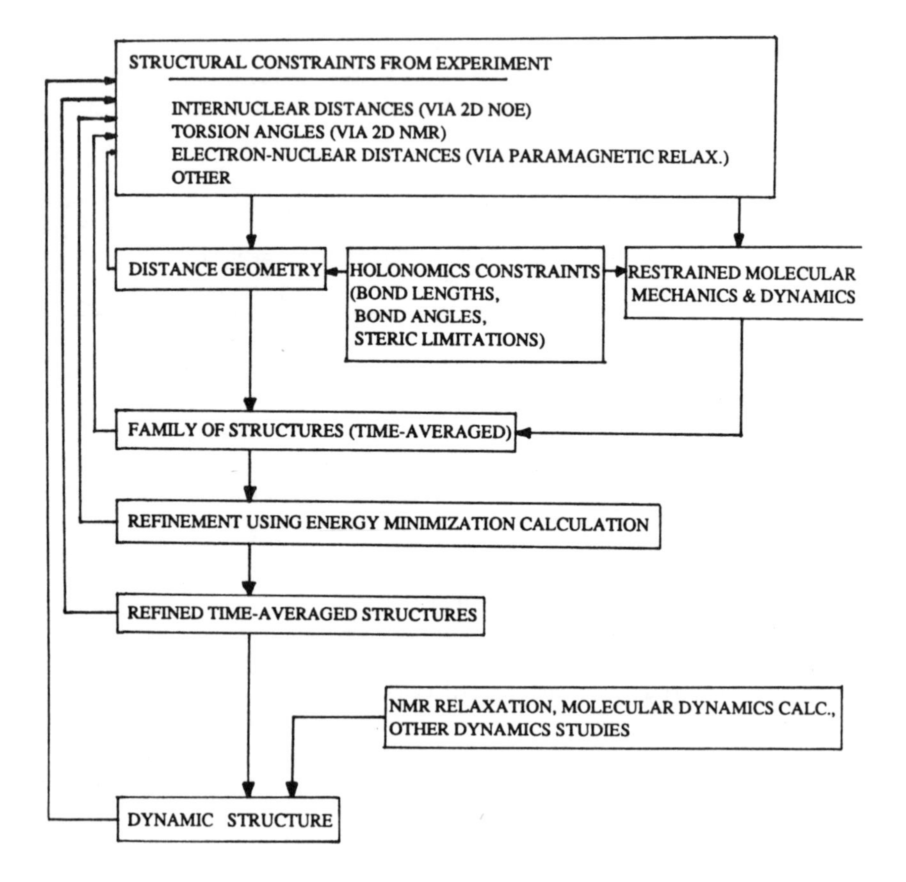

FIGURE 1. Scheme for deriving molecular structures in non-crystalline environments using experimental NMR data in conjunction with computational procedures.

DNA helix which are sequence-dependent and, consequently, guide protein or drug recognition. These subtle variations demand detailed knowledge of the structure and, therefore, accurate internuclear distances. But one can probably define a protein tertiary structure with moderate accuracy using DG or restrained MD calculations without accurately determining interproton distances. A qualitative assessment of the 2D NOE spectrum often suffices to obtain the information necessary for calculation of a modestly high-resolution protein structure in solution. But, in proteins possessing less common structural features, it may be valuable to have more accurate interproton distances. And we will want better defined structures at ligand binding sites (with and without ligand bound). Use of a complete relaxation matrix (CORMA) approach to ascertain interproton distances from 2D NOE peak intensities offers the opportunity of determining protein solution structure with greater accuracy and resolution. Here we will compare different methods of analyzing 2D NOE spectra for internuclear distance and structural information. The most effective techniques employ an iterative method entailing CORMA in concert with molecular mechanics, molecular dynamics or distance geometry calculations.

The 2D NOE experiment was first described and given a firm theoretical foundation in Ernst's laboratory (6,7). We previously described a matrix method for the explicit calculation of the homonuclear 2D NOE intensities expected for a known molecular structure (8) and compared results so obtained for a rigid test molecule (proflavin) in solution with its x-ray crystal structure (9). The essential concepts described in these seminal papers have been further elaborated by several other laboratories (10-14). We call this general approach to the calculation and interpretation of 2D NOE intensities CORMA for COmplete Relaxation Matrix Analysis. The critical feature in CORMA is the explicit treatment of the complete relaxation network in calculating the 2D NOE intensities.

RELATIONSHIP OF INTERPROTON DISTANCES AND 2D NOE INTENSITIES

The effect of cross-relaxation between two protons during the mixing time period τ_m of the 2D NOE experiment is to transfer magnetization between them (7):

$$\mathbf{M}(\tau_m) = \mathbf{a}(\tau_m)\mathbf{M}(0) = e^{-\mathbf{R}\tau_m}\mathbf{M}(0) \tag{1}$$

In Eqn. 1, $\mathbf{M}$ is the magnetization vector describing the deviation from thermal equilibrium ($\mathbf{M} = \mathbf{M}_z - \mathbf{M}_0$), and $\mathbf{R}$ is the matrix describing the complete dipole-dipole relaxation network, where diagonal and off-diagonal elements are

$$R_{ii} = 2(n_i - 1)(W_1^{ii} + W_2^{ii}) + \sum_{j \neq i} n_j(W_0^{ij} + 2W_1^{ij} + W_2^{ij}) + R_{1i};\ R_{ij} = n_i(W_2^{ij} - W_0^{ij}) \tag{2}$$

Here n_i is the number of equivalent spins in a group such as a methyl rotor, and the zero, single, and double transition probabilities W_n^{ij} are given (for isotropic random reorientation of the molecule) by :

$$W_0^{ij} = \frac{q\tau_c}{r_{ij}^6};\ W_1^{ij} = 1.5\frac{q\tau_c}{r_{ij}^6}\frac{1}{1+(\omega\tau_c)^2};\ W_2^{ij} = 6\frac{q\tau_c}{r_{ij}^6}\frac{1}{1+4(\omega\tau_c)^2} \tag{3}$$

where $q = 0.1\gamma^4\hbar^2$. The term R_{1i} represents external sources of relaxation such as paramagnetic impurities and is generally ignored.

a is the matrix of mixing coefficients which *are* proportional to the 2D NOE intensities. This matrix of mixing coefficients is what we wish to evaluate. The exponential dependence of the mixing coefficients on the cross-relaxation rates complicates the calculation of intensities (or the distances). Note that the expression for the above rate matrix is actually still an approximation in that it neglects cross-correlation terms between separate pairwise and higher order interactions (14-15) and, the expressions given above also do not account for second-order effects due to strong scalar coupling (12,16). However, the magnitude of error due to neglect of these effects is small.

METHODS OF ANALYSIS OF 2D NOE SPECTRA

We have compared a few different methods of analyzing 2D NOE spectra for internuclear distance and structural information (17). The analysis generally entails use of hypothetical data sets. This use of hypothetical data is a necessity, since we must know the structure and molecular dynamics exactly in order to understand the effects of any random or systematic errors in experimental spectral intensities or the limitations of the different methods being developed to determine structure. We can calculate the theoretical 2D NOE spectrum for the hypothetical structure using any motional model (8). We can add random (or systematic) noise at any level desired. And we can consider any number of peaks to be overlapping. Furthermore, we can compare the various methods proposed in their abilities to handle realistic spectral limitations (17). In other words, for a given dynamic structure we can create spectra with various realistic problems. Then we can see how well we are able to deduce the structure using the different methodologies without using our *a priori* knowledge of the structure.

Isolated Spin Pair Approximation (ISPA)

The exponential expression of Eqn. 1 can be expanded and the series truncated after the linear term for short ($\tau_m \rightarrow 0$) mixing times. In some studies, NOE build-up curves are obtained to assess whether or not the short mixing time condition is achieved. The practical application of ISPA usually goes one step further to eliminate the dependence on correlation time by scaling all the distances with respect to a known reference distance which is assumed to have the same correlation time as the proton-proton pair of interest. The distance r_{ij} between nuclei i and j is usually determined by comparing the cross-peak intensity a_{ij} with that of a reference cross-peak (a_{ref}) which originates from two protons whose internuclear distance r_{ref} is known (correlation time of ij pair and reference pair assumed to be equal). Then distances are calculated according to:

$$r_{ij} = r_{ref}\left(\frac{a_{ref}}{a_{ij}}\right)^{1/6}$$

Clearly, the chief advantage of using ISPA lies in its simplicity.

But what are the disadvantages of ISPA? Use of short mixing times definitely limits the signal-to-noise ratio obtainable with cross-peaks in the 2D NOE spectrum. One might also consider how reliable are the distances calculated using ISPA? This question was examined (17). It was found that whenever (at least) one proton approaches the "isolated" pair (or both) at a distance ≤ the distance between the pair, the approximation breaks down for practical values of τ_m. Note that on average there are 3.4 protons within 3 Å of any proton in DNA and 4.7 in proteins. The problem becomes worse as molecular motions slow, but is significant with a correlation time of 1 ns. ISPA underestimates distance if $r > r_{ref}$ and overestimates distance if $r < r_{ref}$.

The implication of using ISPA-derived distances with distance geometry is that the bounds, in particular the upper bound, may need to be relaxed more than has been the case in calculations reported. There are also implications for restrained molecular dynamics calculations. If distances are systematically under-estimated, the protein may never be able to get to the vicinity of the global minimum during the MD simulation.

Complete Relaxation Matrix Analysis (CORMA)

A more rigorous method of calculating intensities is to take advantage of linear algebra and the simplifications which arise from working with the characteristic eigenvalues and eigenvectors of a matrix. The rate matrix **R** can be represented by a product of matrices: $\mathbf{R} = \boldsymbol{\chi}\boldsymbol{\lambda}\boldsymbol{\chi}^{T}$ where $\boldsymbol{\chi}$ is the unitary matrix of orthonormal eigenvectors ($\boldsymbol{\chi}^{-1} = \boldsymbol{\chi}^{T}$), and $\boldsymbol{\lambda}$ is the diagonal matrix of eigenvalues. Since $\boldsymbol{\lambda}$ is diagonal, the series expansion for its exponential (and consequently that of the mixing coefficient matrix) collapses:

$$\mathbf{a} = \boldsymbol{\chi} e^{-\boldsymbol{\lambda}\tau_m} \boldsymbol{\chi}^{T} \tag{4}$$

This calculation allows one to readily calculate all the cross-peak intensities for a proposed structural model. Comparison between calculated and measured intensities allows a determination as to the validity of a model structure. We have developed a program for performing this calculation, named CORMA (8,17). Typically, CORMA is used to calculate the 2D NOE peak intensities for a feasible structural model and to compare the calculated intensities with experimental intensities numerically.

We have carried out a number of theoretical calculations as well as experimental studies with CORMA. In summary (8,9,17): Distances up to 6Å (and possibly 7Å) with an accuracy of 10% could be attainable with knowledge of individual relaxation times. Detailed knowledge of the molecular motions is not required for this distance accuracy; it is generally sufficient for a single effective isotropic correlation time to be used in the spectral density expression for any nucleus. The CORMA approach enables an indication of regions of good (or bad) fit between the model and true solution structure. The limitation of CORMA is that one must have a structural model or initial list of distance estimates. One can iteratively fit the simple 2D NOE spectra of small molecules (9); that was initially not possible for larger molecules but has more recently been done (see below). Over the course of our studies, the sophistication of analysis via CORMA developed from evaluation of a few models on the basis of selected intensities to a more detailed analysis of many closely related models using all well-resolved intensities, and finally to iterative fitting of spectra (18-21).

Direct Calculation of Distances from Experimental Spectra (DIRECT)

An obvious solution to the problem of limited (and biased) trial structures is to apply the same computational techniques used in CORMA — but in reverse — to the direct calculation of proton-proton distances from the experimental intensities. This approach has been discussed elsewhere (10,17). Rearrangement of Eqn. 1 gives:

$$\frac{-\ln\left[\dfrac{a(\tau_m)}{a(0)}\right]}{\tau_m} = \mathbf{R} \tag{5}$$

So, assuming isotropic tumbling and using Eqns. 2 and 3, distances can be calculated directly from the rate matrix via diagonalization of the experimental spectrum (matrix).

We have tested the performance of the direct calculation of distances (DIRECT method) in the presence of several different types of data errors: random noise, cross-peak overlap, and diagonal peak overlap (17). For relatively small molecules yielding spectra with very high signal-to-noise, in which most of the major peaks can be resolved and accurately estimated, this is clearly the ideal method of distance determination. However, low resolution (peak overlap) and low (i.e., generally realistic protein or nucleic acid spectral) signal-to-noise hamper the accurate estimation of distances (17). These complications lead to a lack of knowledge about relaxation pathways. This problem is exacerbated by spin diffusion which occurs at longer mixing times. Under these circumstances, a significant amount of spin magnetization is distributed among a large number of cross-peaks, although the amount each may have could be so small that the intensity is less than the signal-to-noise ratio. Consequently, the DIRECT approach has the same limitations as ISPA, namely, short mixing times are better, and longer distances will be underestimated. Regardless, the DIRECT method is distinctly better than ISPA.

Refinement of Interproton Distances using COMATOSE

Limitations of the ISPA and DIRECT methods for accurately determining interproton distances led us to pursue the less direct and more time-consuming approach of iterative least-squares refinement of structure based on the 2D NOE intensities. We have developed the program COMATOSE (COmplete Matrix Analysis Torsion Optimized StructurE) for refinement of molecular structure based on 2D NOE intensities (17). The goal is to optimize a trial structure while minimizing the error between calculated and observed intensities. We have investigated the effects of experimental errors in peak intensities and peak overlap on the refinement process. Use of Cartesian coordinates to define proton positions provided too many parameters for refinement to proceed properly. To reduce the number, one can take advantage of holonomic structural constraints and utilize internal coordinates.

We have found COMATOSE to work reasonably well for DNA fragments, even for a fairly long mixing time of 250 ms with experimentally realistic random noise and overlapping peaks. The other methods (ISPA,DIRECT) compared were not very good at such a long mixing time. Consequently, it is feasible that sufficient cross-peak intensity can be measured that longer distances can be determined. Not only does COMATOSE improve the distances, but the spectral density can be refined simultaneously. In short, the COMATOSE refinement process is much more cpu time-consuming than simpler methods for estimating the distances from 2D NOE intensities such as ISPA or the DIRECT method. However, it does not suffer from the systematic errors associated with such methods. No information is needed from the diagonal peak intensities (which

Matrix **A**nalysis of **R**elaxation for **DI**scerning **G**eometR**y** of an **A**queous **S**tructure
(MARDIGRAS)

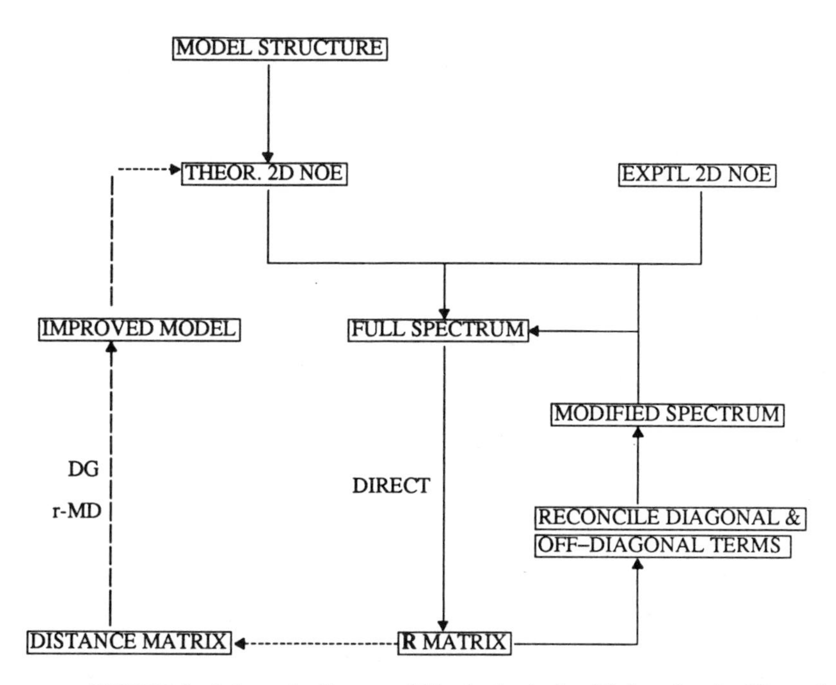

FIGURE 2. Schematic diagram of Matrix Analysis of Relaxation for DIscerning GeometRy of an Aqueous Structure (MARDIGRAS). A model structure is used to generate a theoretical 2D NOE spectrum (using CORMA). Wherever possible, experimental intensities are substituted into the theoretical spectrum to yield a full spectrum suitable for direct solution of the rate and distance matrices. This process incorporates all the effects of network relaxation and spin diffusion and will be relatively accurate, depending on how close the model is to the true structure. An augmented matrix or modified spectrum utilizes known distance constraints and required internal consistancy between diagonal and off-diagonal elements. To improve the structure, distance geometry (DG) or restrained molecular dynamics (r-MD) can be used to generate an improved model which incorporates the distance restraints generated by the previous pass of the intensity/distance calculation.

are typically overlapping and difficult to measure), and overlapping or tiny cross-peaks pose no problem (other than loss of potential additional information) since these intensities are simply not included in the calculation of the error to be minimized. Major limitations of the COMATOSE approach are (a) the neglect of energetic and steric constraints during the structure refinement process and (b) ability of COMATOSE to find a local minimum rather than a global minimum. These occasionally allow conformations to be generated which satisfy the distance constraints imposed by the NOE intensities, but which are otherwise unacceptable. It may be possible to overcome both of these limitations by employing COMATOSE in concert with molecular mechanics or molecular dynamics calculations.

It is obvious that an initial structure is necessary for refinement via COMATOSE. In the case of DNA fragments, it is reasonable to start with standard B form DNA and refine that structure using the experimental 2D NOE spectra. In the case of proteins, it may be appropriate to obtain an initial structure using ISPA (or, better, DIRECT) via analysis of the distances with either the methods of distance geometry or restrained molecular dynamics. COMATOSE could then be used to refine that structure.

Matrix Analysis of Relaxation for DIscerning GeometRy of an Aqueous Structure (MARDIGRAS)

A potentially quite useful extension of the DIRECT calculation of distances is iterative relaxation matrix analysis (IRMA) proposed by Prof. Robert Kaptein. We have developed a variant of IRMA termed MARDIGRAS. The general scheme is shown in Figure 2. The critical feature of MARDIGRAS (or IRMA) is the generation of the augmented intensity matrix which contains all the scaled experimental intensities and the intensities calculated from a suitable model structure (obtained from model building, distance geometry, etc.). By solving the augmented intensity matrix for distances, one generates a distance set which will be in reasonable agreement with the model, but which is also partially restrained by experimental intensities. By iterating through the cycle of structure generation via DG or MD on one branch and CORMA-type calculations on the other, one will eventually reach a self-consistent structure which incorporates all structural information inherent in the 2D NOE intensities. The variation introduced in MARDIGRAS, and shown on the lower right in Figure 2, utilizes requirements for internal consistency in the relaxation matrix itself rather than iterating through the computer-time consuming restrained MD or DG procedures after a single pass through the relaxation matrix. With MARDIGRAS, constrained distances are used to give relaxation matrix intensities as a first approximation, and diagonal and off-diagonal elements are required to be consistent. This yields a modified spectrum which is then re-combined with the experimental 2D NOE data to yield a new augmented matrix.

We have made some simple tests of the accuracy of the distances derived from the augmented intensity matrix. The idealized structure was based on x-ray coordinates of bovine pancreatic trypsin inhibitor (BPTI) to which we appropriately added protons. CORMA, with a realistic random noise level of ± 0.003, was used to generate the "experimental" spectrum. For the initial model structure, we generated a random coil structure based on the BPTI primary sequence. As seen in Figure 3, initial distances varied considerably from ideal x-ray distances. But with successive cycles through MARDIGRAS, distances improved dramatically. The resulting distances could then be used as either distance geometry or molecular dynamics input.

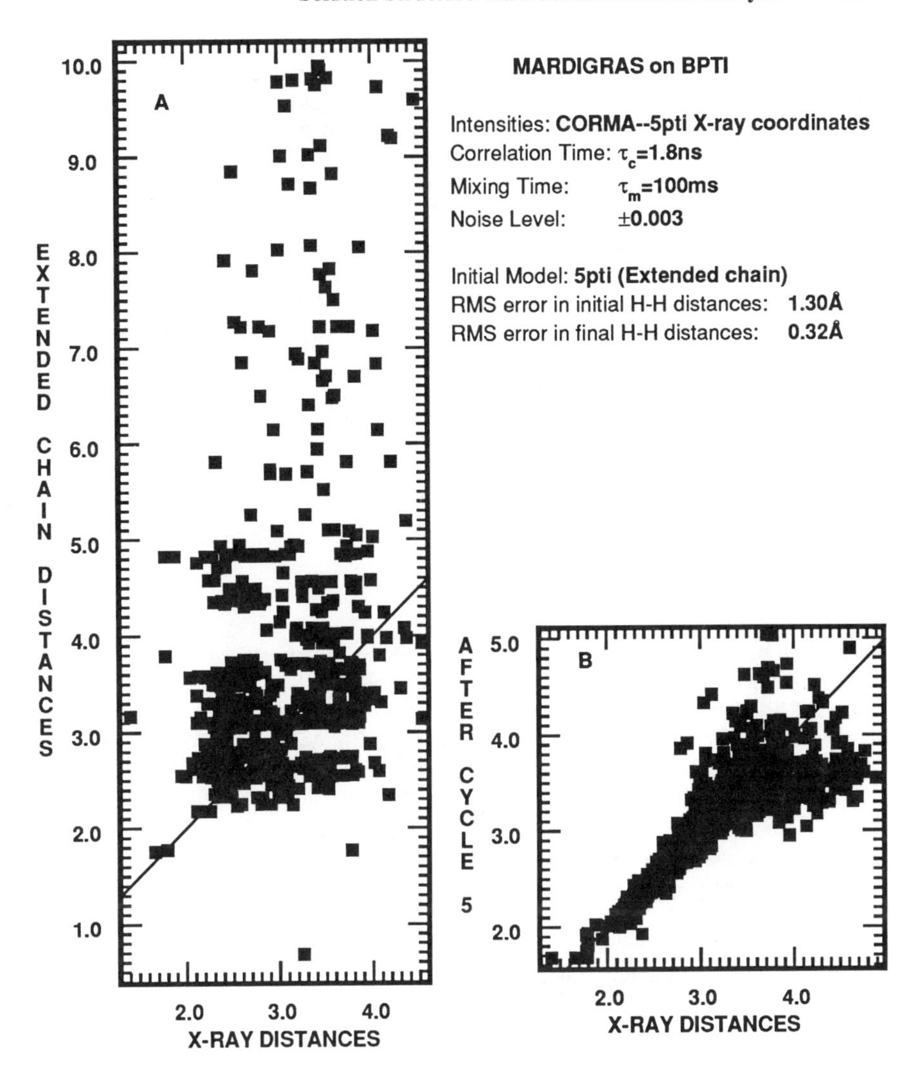

FIGURE 3. Comparison of initial interproton distances from random coil BPTI sequence with those of the "true" x-ray structure, and comparison of the interproton distances following MARDIGRAS refinement with those of the "true" x-ray structure.

CONCLUDING REMARKS

The original scheme presented for structure determination (Figure 1) indicated both distance geometry and restrained molecular dynamics could be usefully employed. Both should be used in conjunction with the complete relaxation matrix analysis of the

2D NOE data. It is not possible to know now which, if any, method will prove superior and will be the method of choice five years from now. In fact, it may induce greater confidence to derive essentially the same structure using a couple of different approaches. An important aspect of this development is that we can always compare theoretical spectra calculated from any interim refinement model with experimental spectra. (Notice the "feedback" arrows in Figure 1.)

We might consider what criteria will be used to assess the structures during refinement and for comparison between structures obtained using different refinement strategies. First, it should be noted that 2D NOE peak intensities are so sensitive to interproton distances that it is very easy to create plausible model structures which yield theoretical spectra at strong variance with the experimental spectra. We have in the past used a difference index as a means of comparing between closely related model structures (20,21). However, one can use a residual index, in keeping with the practice of crystallographers, which essentially normalizes our previous difference index. The residual index can be defined as

$$R_1 = \frac{\sum_i (I_o^i - I_c^i)}{\sum_i (I_o^i)} \tag{6}$$

where the summation is over all protons for the observed and calculated intensities. An alternative definition of the R value, more in keeping with the root-mean-square deviation, has also been given (22). Methods of assessing meaningful changes in the R value when the model is altered are also straight-forward (22). As any crystallographer would acknowledge, although the R value is not a perfect guide, it is a useful one. For x-ray crystallography, there is a maximum value of R_1, 0.83, for a random distribution of atoms. Lower values imply a better match. X-ray crystallographers have, through experience, managed to correlate numerical values to an assessment of the "goodness" of any model's fit to the experimental data. The actual numerical values for R which correlate to a particular quality of fit are expected to differ from that of the crystallographers, but we should be able to use R values (or difference index values) in the same way as a measure of solution structure refinement.

References

1. Havel, TF, Kuntz ID, Crippen GM, (1983) ''The Theory and Practice of Distance Geometry,'' *Bull Math Biol* **45**, 665-720.
2. Havel, TF, Wuthrich K, (1985) ''An Evaluation of the Combined Use of Nuclear Magnetic Resonance and Distance Geometry for the Determination of Protein Conformations in Solution,'' *J Mol Biol* **182**, 281-294.
3. Singh UC, Weiner PK, Caldwell J, Kollman PA, '','' AMBER 3.0, University of California, San Francisco (1986).
4. Karplus, M, McCammon JA, (1979) ''Protein structural fluctuations during a period of 100 ps,'' *Nature* **277**, 578.

5. Gunsteren WF van, Boelens R, Kaptein R, Zuiderweg ERP, in *Nucleic Acid Conformation and Dynamics, Report of NATO/CECAM Workshop*, ed. W.K. Olson (1983).
6. Jeener, J, Meier BH, Bachmann P, Ernst RR, (1979) "Investigation of Exchange Processes by 2D NMR Spectroscopy," *J Chem Phys* **371**, 4546-4553.
7. Macura, S, Ernst RR, (1980) "Elucidation of Cross Relaxation in Liquids by 2D NMR Spectroscopy," *Mol Phys* **41**, 95-117.
8. Keepers, JW, James TL, (1984) "A Theoretical Study of Distance Determinations from NMR. Two-Dimensional Nuclear Overhauser Effect Spectra," *J Magn Reson* **57**, 404-426.
9. Young, GB, James TL, (1984) "Determination of Molecular Structure in Solution *via* Two-Dimensional Nuclear Overhauser Effect Experiments: Proflavine as a Rigid Molecule Test Case," *J Am Chem Soc* **106**, 7986-7988.
10. Olejniczak, ET, Gampe Jr RT, Fesik SW, (1986) "Accounting for Spin Diffusion in the Analysis of 2D NOE Data," *J Magn Reson* **67**, 28-41.
11. Macura, S, Farmer BT, Brown LR, (1986) "An Improved Method for the Determination of Cross-Relaxation Rates From NOE Data," *J Magn Reson* **70**, 493-499.
12. Kay, LE, Scarsdale JN, Hare DR, Prestegard JH, (1986) "Simulation of Two-Dimensional Cross-Relaxation Spectra in Strongly Coupled Spin Systems," *J Magn Reson* **68**, 515-525.
13. Massefski Jr, W, Bolton PH, (1985) "Quantitative Analysis of Nuclear Overhauser Effects," *J Magn Reson* **65**, 526-530.
14. Bull, TE, (1987) "Cross-Correlation and 2D NOE Spectra," *J Magn Reson* **72**, 397-413.
15. Werbelow L, Grant DM, (1978) *Adv Magn Reson* **9**, 189.
16. Kay, LE, Holak TA, Johnson BA, Armitage IM, Prestegard JH, (1986) "Second Order Effects in Two Dimensional Cross Relaxation Sprctra of Proteins : Investigation of Glycine Spin Systems," *J Am Chem Soc* **108**, 4242-.
17. Borgias, BA, James TL, (1988) "COMATOSE: A Method for Constrained Refinement of Macromolecular Structure Based on Two-Dimensional Nuclear Overhauser Effect Spectra," *J Magn Reson* **79**, 493-512.
18. Broido, MS, James TL, Zon G, Keepers JW, (1985) "Investigation of the Solution Structure of a DNA Octamer $[d\text{-}(GGAATTCC)]_2$ Using Two-Dimensional Nuclear Overhauser Effect Spectroscopy," *Eur J Biochem* **150**, 117-128.
19. Jamin, N, James TL, Zon G, (1985) "Two-Dimensional Nuclear Overhauser Effect Investigation of the Solution Structure and Dynamics of the DNA Octamer $[d\text{-}(GGTATACC)]_2$," *Eur J Biochem* **152**, 157-166.
20. Suzuki, E.-I., Pattabiraman N, Zon G, James TL, (1986) "Solution Structure of $[d\text{-}(AT)_5]_2$ *via* Complete Relaxation Matrix Analysis of 2D NOE Spectra and Molecular Mechanics Calculations," *Biochemistry* **25**, 6854-6865.
21. Zhou, N, Bianucci AM, Pattabiraman N, James TL, (1987) "Solution Structure of $[d\text{-}(GGTATACC)]_2$: Wrinkled D Structure of the TATA Moiety," *Biochemistry* **26**, 7905-7913.
22. Hamilton, WC, (1964) *Statistics in Physical Science*, Ronald Press, New York.

Frontiers of NMR in Molecular Biology, pages 189-194

DETERMINATION OF PROTEIN STRUCTURE IN AQUEOUS SOLUTION WITH AND WITHOUT DISTANCE CONSTRAINTS[1]

Harold A. Scheraga

Baker Laboratory of Chemistry, Cornell University
Ithaca, New York 14853-1301

ABSTRACT Two-dimensional NMR experiments on proteins in aqueous solution provide information about distances in the range of 2-5 Å, and non-radiative energy transfer (fluorescence) experiments provide similar information about distances in the range of 20-40 Å. Information from both ranges of distances is the most effective in defining a structure. The deviations among such closely related structures can be reduced by carrying out conformational energy calculations on the family of structures obtained from such experiments. In the absence of such experimental information about distance constraints, conformational energy calculations (with available procedures to surmount the multiple-minima problem) can lead to the three-dimensional structure of a protein in aqueous solution.

INTRODUCTION

The three-dimensional structures of proteins in aqueous solution can be determined by acquiring a sufficient number of distance constraints or by conformational energy calculations. Distance constraints can be obtained from a combination of protein modification and physical chemical measurements (1), from 2D NMR experiments (2), or from non-radiative energy transfer experiments (3). The distance constraints are then processed by distance geometry algorithms (4,5) to obtain a family of closely-related structures.

---

[1]This work was supported by a research grant from the National Institute of General Medical Sciences (GM-24893) of the National Institutes of Health.

This family of structures can be narrowed down to a unique structure by energy minimization. Alternatively, conformational energy calculations (without distance-constraint information from NMR and fluorescence experiments) can lead to stable conformations of polypeptides, proteins, and protein complexes (6). This paper is concerned with an assessment of the information that is provided by distance constraints for pairs of residues in proteins, and with some computational aspects of theoretical approaches to determine protein structure in solution.

INFORMATION FROM DISTANCE CONSTRAINTS

We consider here the question as to the kind, the number, and the accuracy needed for the distance constraints in order to determine the conformation of a protein to any given degree of accuracy. To answer this question, we make use of a relationship that provides a measure of the r.m.s. deviation between a computed and a known structure, when the computed structure is generated under a given set of distance constraints. Bovine pancreatic trypsin inhibitor (BPTI), represented as a set of N (= 58) C^{α} atoms, will be used an an example, and further details about the information summarized here can be found elsewhere (7,8).

If d_{ij} is the distance between C_i^{α} and C_j^{α}, then the conformation of a protein of N points can be determined uniquely if any of the following three sets of data are known exactly:

a. 3N Cartesian coordinates (actually 3N-6 degrees of freedom, the remaining 6 serving only to fix the translational and rotational positions of the whole molecule).
b. 3N-6 variables, distributed among N-1 virtual bond lengths (C_i^{α} - to - C_{i+1}^{α}), N-2 virtual bond angles, and N-3 dihedral angles around the virtual bonds.
c. 3N-6 variables, distributed among N-1 values of $d_{i,i+1}$, N-2 values of $d_{i,i+2}$, and N-3 values of $d_{i,i+3}$ (an additional N-4 values of $d_{i,i+4}$ must be known, but only approximately, to define the sign of $d_{i,i+3}$). Alternatively, a set of any other 4N-10 distances may be used.

In actual practice, however, it is usually not possible to obtain this number of distances experimentally and, furthermore, the distances are not known exactly.

To assess the quality of the information that can be expected to be obtainable from experiment, we introduce a quantity I_{ij}, where $1-I_{ij}$ is a measure of the ambiguity in our knowledge of d_{ij}; I_{ij} is computable from the statistics of chain molecules (9). Assuming that the I_{ij}'s are independent, the mean ambiguity, H, in the conformation of the whole molecule is then

$$H = 1 - \frac{1}{N(N-1)/2} \sum_{i<j} I_{ij} \tag{1}$$

where N(N-1)/2 is the total number of pairs. The quantity H is related to the r.m.s. deviation of the generated conformation from the native one, with the generated conformation being consistent with assumed upper and lower bounds, u_{ij} and l_{ij}, respectively on d_{ij}, which determine I_{ij}.

Using sets of conformations of BPTI generated by Havel et al (10), for which E_x is the mean value of the r.m.s. deviation of each of the structures in any given set from the X-ray crystal structure, Wako and Scheraga (8) calculated values of H from eq. 1 and showed that the data could be fit well by the equation

$$E_x = 10.4\ H^{2.53} \tag{2}$$

Equation 2 provides a relationship between the ambiguity H and the r.m.s. deviation E_x.

Values of I_{ij} (and hence of H from eq. 1) were computed for different numbers of pairs of residues, for assumed values of u_{ij} and l_{ij}. Values of E_x were then obtained from these values of H by means of eq. 2. Plots of E_x against the number of pairs of residues revealed a number of interesting conclusions (8). For example, if $|i-j|$ is the "distance" between residues i and j along the amino acid sequence, then, to obtain the computed conformation of BPTI with an r.m.s. deviation of less than 2 Å from the native conformation, the values of d_{ij} of more than ~80 pairs (half of them with $5 \leq |i - j| \leq 20$ and the other half with $21 \leq |i - j| \leq 57$) must be known exactly, or of more than ~150 pairs (half of them with $5 \leq |i - j| \leq 20$ and the other half with $21 \leq |i - j| \leq 57$) must be known with an error no greater than ~2 Å; alternatively, the same r.m.s. deviation of less than 2 Å from the native structure can be achieved by the computed conformation if more than ~160 pairs are chosen so that 20 Å is assigned as the lower limit for half of these d_{ij}'s (for those pairs in the native protein that are separated by $\geq$20 Å) and 10 Å is assigned as the upper limit for the other half of these d_{ij}'s (for those pairs in the native protein that are separated by $\leq$10 Å). In all of the above examples, all values of $d_{i,i+1}$ were fixed at 3.8 Å and all values of $d_{i,i+2}$ were confined to the range 4.5-7.2 Å (the minimum and maximum possible values for a polypeptide chain). For a given number of constraints, information about pairs with large $|i-j|$ or small d_{ij} is more effective in determining the conformation than is information about pairs with small $|i-j|$ or large d_{ij}. It is found, however, that information that includes both small and large $|i-j|$ or both small and large d_{ij} is the most effective. Two-dimensional NMR experiments provide data for small d_{ij} (2) whereas non-radiative energy transfer experiments provide data for large d_{ij} (3). Thus far, the two techniques have not been used together to determine the structure of a protein. Using one of these techniques alone, e.g. 2D NMR spectroscopy, one can obtain only a family of related structures [see Fig. 2 of reference 11 for a family of structures of murine epidermal growth factor (EGF) obtained by applying a distance geometry algorithm to the distances deduced from 2D NMR experiments]. A unique structure of EGF is obtainable from this family by application of energy-minimization techniques; such calculations are in progress.

CONFORMATIONAL ENERGY CALCULATIONS

When no distance constraints are available from experiment, then reliance is placed on conformational energy calculations (6). In order to carry out such calculations, it is necessary to have (a) reliable potential functions, (b) procedures for local minimization (or efficient Monte Carlo searches), and (c) methods to surmount the multiple-minima problem. The available potential functions and minimization procedures are adequate enough to yield results on a variety of systems that have been checked by experiment (1,6), and procedures are now available to surmount the multiple-minima problem (6). These include:

A. Build-up method
B. Build-up with limited constraints
C. Calculations with constraints
D. Use of homology
E. Optimization of electrostatics
F. Monte Carlo-plus-minimization
G. Electrostatically driven Monte Carlo
H. Adaptive importance sampling Monte Carlo
I. Increase in dimensionality
J. Pattern-recognition-importance sampling-minimization
K. Deformation of the potential energy hypersurface.

A descriptive summary of these procedures is provided in reference 6. They are used separately, and in various combinations with each other, to locate the approximate native conformation of a globular protein. They are all intended as the _initial_ approaches in the computations. In the _final_ stages, the results from all of these procedures are collated into an approximate three-dimensional structure whose energy should lie in the potential well containing the global minimum (i.e. this structure should be a good approximation of the native structure). Then, the complete conformational energy of this structure is minimized, taking _all_ pairwise interactions (over the whole molecule) into account.

SUMMARY

Conformational energy calculations provide the basis for determining the three-dimensional structures of proteins in aqueous solution. Two-dimensional NMR experiments lead to distance constraints (up to 5 Å) which, with the aid of distance geometry, provide a family of similar structures. Longer-range distance constraints (20-40 Å), obtainable from non-radiative energy transfer experiments, serve to reduce the deviations among the calculated family of structures. Conformational energy calculations can then lead to a unique structure. In the absence of distance constraints from NMR and fluorescence experiments, methodology is now available to surmount the multiple-minima problem so that conformational energy calculations can lead directly to the three-dimensional structure of a protein in aqueous solution.

REFERENCES

1. Scheraga HA (1984) Carlsberg Res Commun 49:1.
2. Braun W, Bösch C, Brown LR, Gō N, Wüthrich K (1981) Biochim Biophys Acta 667:377.
3. McWherter CA, Haas E, Leed AR, Scheraga HA (1986) Biochemistry 25:1951.
4. Crippen GM (1981). "Distance Geometry and Conformational Calculations." Chichester: Research Studies Press.
5. Braun W, Gō N (1985) J Mol Biol 186:611.
6. Scheraga HA (1989) Chemica Scripta, in press.
7. Crippen GM (1977) J Comput Phys 24:96.
8. Wako H, Scheraga HA (1981) Macromolecules 14:961.
9. Flory PJ (1953). "Principles of Polymer Chemistry." Ithaca, NY: Cornell University Press, p. 407.
10. Havel TF, Crippen GM, Kuntz ID (1979) Biopolymers 18:73.
11. Montelione GT, Wüthrich K, Nice EC, Burgess AW, Scheraga, HA (1987) Proc Natl Acad Sci USA 84:5226.

Frontiers of NMR in Molecular Biology, pages 195-213

N.M.R. OF COMPLEX CARBOHYDRATES*

Herman van Halbeek

Complex Carbohydrate Research Center, University of Georgia, Athens, GA 30613, USA

Abstract —— An integrated approach of homonuclear (COSY, HOHAHA) and ^{1}H-detected heteronuclear (HMQC, HMBC) shift-correlation NMR experiments for the sequencing of complex carbohydrates is presented. The method is illustrated for a heptasaccharide obtained from swiftlet salivary mucus glycoproteins and for a decasaccharide derived from the xyloglucan of rapeseed hulls. The presented strategy is also applicable for the purpose of locating the sites of naturally occurring *O*-acyl groups in a complex carbohydrate. This aspect of the method is demonstrated for a highly esterified trisaccharide isolated from the stem bark of *Mezzettia leptopoda (Annonaceae)*.

INTRODUCTION

Complex carbohydrates are active as chemical messengers both in animals and plants. The oligosaccharide chains of animal glycoproteins and glycolipids are known to play a role as tissue-specific cell-surface antigens, as receptors for hormones, toxins, bacteria and viruses, in targeting proteins to subcellular organelles, in cell differentiation, malignant cell transformation and tumor metastatic power [1,2]. Plant cell wall oligo- and polysaccharides are involved in the defense mechanism of plants against microbial infections (*i.e.*, host-pathogen interaction), and in controlling morphogenesis in the

* This research is supported by DOE grant DE-FG09-87ER13810, NIH grant HL-38213 and CFF grant G-169 (8-1 & 9-2).

tissues and cells of plants [3]. In addition, complex carbohydrates have important physicochemical functions. The mucus glycoproteins in animal gastrointestinal and respiratory tracts serve as a protective barrier over the epithelial cell lining [1,2], and polysaccharides convey rigidity to plant cell walls [3].

To gain an insight into the biological and physicochemical functions of complex carbohydrates at the molecular level, knowledge of their complete, *i.e.*, primary and secondary, structures is a prerequisite. The complete structural characterization of complex carbohydrates involves determining *(i)* the type, number and primary sequence of the constituting glycosyl residues (including the occurrence of branch points, and the location of appended non-carbohydrate groups such as *O*-alkyl, *O*-acyl, *O*-phosphate and *O*-sulfate groups), and *(ii)* the three-dimensional conformation(s) and dynamics in solution. Although determining the complete structure of a carbohydrate usually requires the application of a combination of chemical, enzymic, and spectrometric methods, NMR spectroscopy is the single most powerful technique for the accomplishment of this task.

The nuclei of interest in the NMR studies of carbohydrates are predominantly ^{1}H and ^{13}C. The natural allocation of hydrogen and carbon atoms in carbohydrate molecules lends itself to reveal their complete structure through ^{1}H- and ^{13}C-NMR analysis. Here we present an NMR approach aimed at solving two crucial problems in carbohydrate primary structural analysis, namely, the elucidation of the sequence of glycosyl residues, and the location of naturally occurring appended *O*-acyl groups.

DETERMINING THE PRIMARY STRUCTURE OF A COMPLEX CARBOHYDRATE BY N.M.R. SPECTROSCOPY

1-D ^{1}H-NMR Spectroscopy —— The first step in the structural analysis of a complex carbohydrate is to obtain an integrated one-dimensional (1-D) ^{1}H-NMR spectrum of the native material dissolved in D_2O or another suitable solvent. The spectrum obtained represents an "identity card" of the carbohydrate. In cases where the spectrum does not match any of the spectra in existing databases (see *e.g.* [4-7]), attempts can be made to interpret the ^{1}H-NMR spectrum in terms of (partial elements of) the primary structure (including anomeric configurations and positions of glycosidic linkages) of the carbohydrate by using the well-documented [7,8] "structural-reporter-group" concept. In outline, the crowded region in the center of the spectrum (between 3 and 4 ppm) is virtually completely neglected, and only the positions and patterns of those signals that are individually observable are examined. Particularly useful structural reporter-groups in such 1-D analyses are (*i*) the anomeric (H1) protons; (*ii*) the protons attached to the carbon atoms in the direct vicinity of a substitution position; (*iii*) the protons attached to deoxy carbon atoms; and (*iv*) methyl protons, *e.g.*, in *N*- and *O*-acetyl groups. Partial or even complete primary structure determination is possible from the 1-D ^{1}H-NMR spectrum provided that structurally related compounds have been previously characterized by ^{1}H-NMR spectroscopy. It is recommended that the glycosyl-residue composition be obtained independently by chemical analysis, and the molecular weight be verified by FAB mass spectrometry. The 1-D ^{1}H-NMR structural-reporter-group approach is extremely useful in those cases where only minute amounts of a pure carbohydrate (in the order of 50-100 μg) are available (see *e.g.* [9-11]).

Homonuclear 2-D NMR Spectroscopy: Sequence-specific Assignments of ^{1}H Resonances by Tracing J*-coupling Networks* —— In those cases where the 1-D ^{1}H-NMR spectrum of the carbohydrate under investigation does not resemble those of previously encountered structures, 2-D NMR analysis is the method of choice, given that sufficient sample is available. Primary structural analysis of complex carbohydrates *via* 2-D NMR is based primarily on tracing through-bond *J*-coupling connectivities (compare [12-15]). Glycosyl residues are identified as homonuclear *J*-coupling (*intra*-ring connectivities) networks (see below); then, glycosidic linkages, and thus the sequence, of the glycosyl residues can be obtained by tracing *inter*-residue connectivities such as homonuclear NOEs or, preferably [16], three-bond, heteronuclear *J*-couplings (see next section).

The strategy for obtaining sequential assignments in ^{1}H-NMR spectra of carbohydrates basically involves two different types of NMR experiments. The various glycosyl residue spin systems of a particular carbohydrate are identified by a combination of 2-D direct (COSY) and relayed (either RELAY or TOCSY/HOHAHA) *J*-correlated spectroscopy. Because of the consecutive allocation of protons around each glycosyl ring, a vicinal *J*-coupled connectivity exists for each residue such that each proton leads to

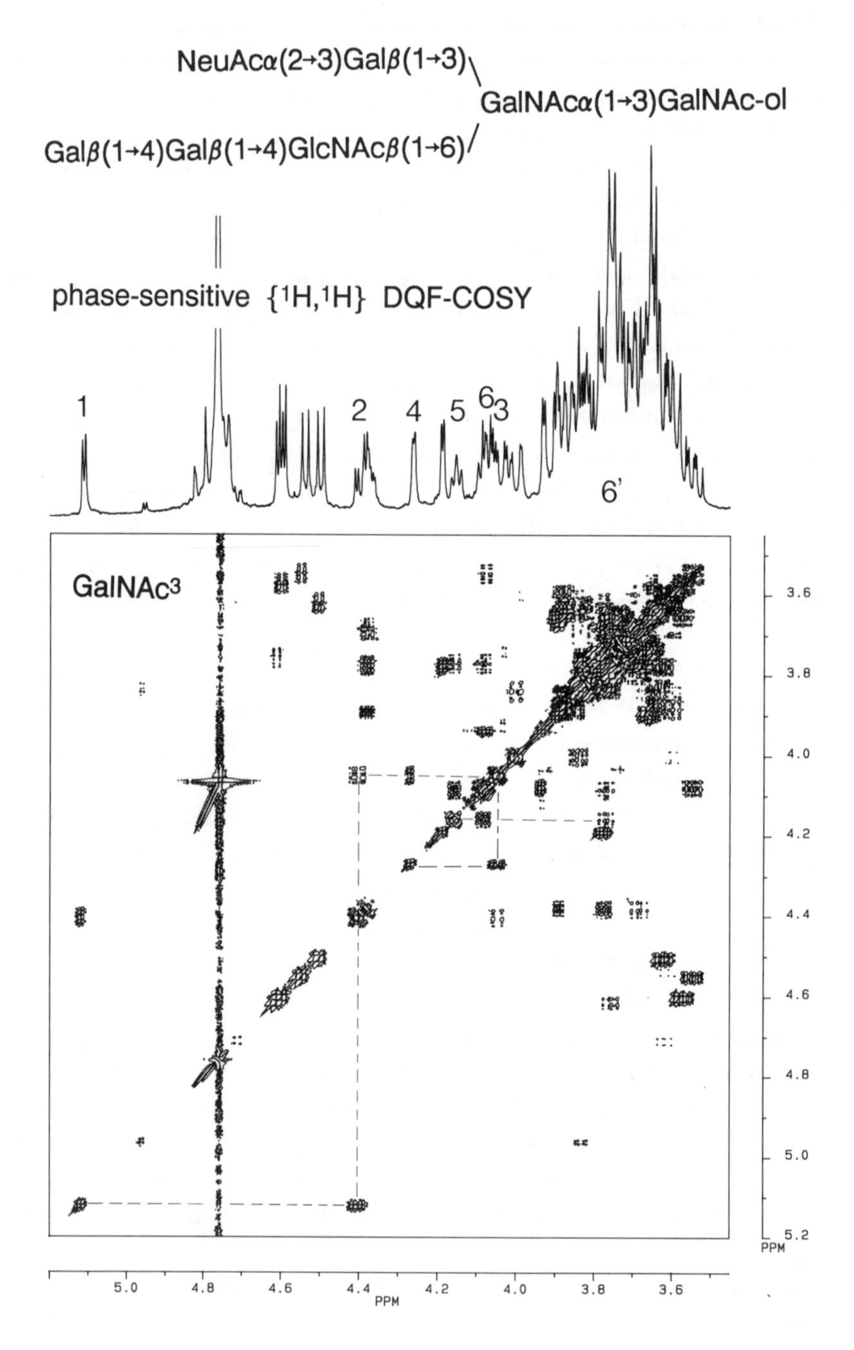
NeuAcα(2→3)Galβ(1→3)
GalNAcα(1→3)GalNAc-ol
Galβ(1→4)Galβ(1→4)GlcNAcβ(1→6)
phase-sensitive {1H,1H} DQF-COSY
1
2
4
5
6
3
6'
GalNAc3
3.6
3.8
4.0
4.2
4.4
4.6
4.8
5.0
5.2
PPM
5.0
4.8
4.6
4.4
4.2
4.0
3.8
3.6
PPM

Figure 1.

the next proton of the ring. Conveniently, the anomeric proton (H1) of each aldosyl residue resonates in a characteristic region ($4.5 < \delta < 6$ ppm) that contains few other signals. Thus the *J*-connectivity trail for such residues can be initiated in this region. The actual magnitudes of the *J*-couplings reveal the stereochemistry and configuration at each carbon; the complete set of *J*-couplings characterizes the identity of the residue.

Fig. 1 shows the COSY spectrum of a heptasaccharide isolated [17] from the salivary mucin glycoproteins (nest cementing substance) of the Chinese swiftlet. The COSY experiment reveals connectivities for each pair of vicinal protons (and also for geminal protons); the *J*-connectivity trail for the GalNAc[3] residue[‡], the only α-linked aldosyl residue in the oligosaccharide, is indicated. Phase-sensitive, pure-absorption COSY was applied to eliminate the dispersion character of the diagonal peaks allowing identification of cross-peaks lying immediately adjacent to the diagonal; also, the resolution within the cross-peaks is improved over the conventional (magnitude-)presentation of COSY data. In addition, double-quantum-filtering (DQF) was applied to remove singlets (single-quantum transitions) from the diagonal and, more importantly, the accompanying t_1 noise from the off-diagonal parts of the spectrum.

The GalNAc[3] residue is exceptional in that it is the only residue which shows essentially its complete *J*-connectivity from H1 to H6 and H6' in the COSY experiment (see Fig. 1). The limitation of the COSY spectrum with respect to the tracing of complete *J*-connectivity trails for residues other than GalNAc[3] in the mucin heptasaccharide is in the overlap of two or more cross-peaks remote from the diagonal at virtually the same chemical shift, due to the limited dispersion of the protons involved (see the $3.5 < \delta$

FIGURE 1. Section of the 2-D $\{^1H,^1H\}$ DQF-COSY spectrum of the heptasaccharide isolated from the salivary mucins of the Chinese swiftlet (*genus Collocalia*). The structure of the compound and the conventional 1-D ^1H spectrum are shown on top of the contour map. Dashed lines connect signals from protons (1 to 6 and 6') of the branching GalNAc[3] residue. The sample contained 4 mg of the heptasaccharide in 0.4 ml D_2O at pD 6 and 25°C. The COSY experiment was performed in the phase-sensitive mode (TPPI) at 500 MHz, on a Bruker AM-500. Data matrix: 512 x 2048; 64 scans per t_1 value. Sine bell windows were used in both dimensions. Total measuring time: 4 hours.

‡ A superscript at the name of a glycosyl residue indicates to which position of the adjacent residue it is linked.

< 4.0 ppm region; Fig.1). This problem may be overcome either by relayed-coherence-transfer (RELAY) spectroscopy [18,19], or by total-correlation-spectroscopy (TOCSY) [20], also known as homonuclear Hartmann-Hahn (HOHAHA) spectroscopy [21]. Both methods are valuable for establishing *J*-connectivities between two remote nuclei within a given spin system (glycosyl residue). However, the coherence transfer process in HOHAHA, taking place in the rotating frame while the magnetization is locked in the transverse plane, is far less dependent upon the *J*-signature of the glycosyl residues than in RELAY spectroscopy. The HOHAHA method has the potential of unraveling the composite ^{1}H-NMR spectrum of a carbohydrate into a subset of spectra derived from individual glycosyl residues. Fig. 2a shows the structure of a decasaccharide isolated from the xyloglucan of rapeseed hulls [22,23], along with the contour map of its 2-D HOHAHA spectrum. As illustrated in Fig. 2b, this technique allowed the generation of complete subspectra (*i.e.*, including correlations between H1 and H6, and H1 and H6') for each glycosyl residue (see also [21,24,25]).

In summary, all ^{1}H chemical shifts (and homonuclear coupling constants) were obtained for both the mucin heptasaccharide and the xyloglucan decamer with the combination of two 2-D homonuclear shift-correlation techniques, namely, COSY and HOHAHA. The resonances were grouped into families of signals belonging to the same network of *J*-coupled spins; thus, subspectra for each constituent glycosyl residue were obtained. The subspectra enabled the identification of each constituent glycosyl residue, and provided the configuration of its glycosidic linkage to the next glycosyl residue. Although the HOHAHA method, in principle, offers the possibility of obtaining all resonance assignments for a carbohydrate in a single experiment, in practice, several HOHAHA experiments differing in spin-lock time ("mixing time") were found to be necessary to trace the magnetization transfer from H1 all the way to H6 and H6' step by step.

Heteronuclear 2-D NMR Spectroscopy: Sequencing of Carbohydrates by Tracing $^3J_{CH}$ *Couplings* —— Inter-ring connectivities cannot be made by $\{^1H,^1H\}$ scalar connectivities since such $^4J_{HH}$ are usually not observed. Therefore, it is necessary to make use of either homonuclear dipolar couplings (NOE measurements), or of the long-range heteronuclear coupling constants $^3J_{CH}$ across the glycosidic linkages. A serious drawback of the first method is that the interglycosidic NOE between H1 and the proton on the attachment site of the other glycosyl residue is not necessarily the largest inter-residue NOE effect [15,27]. NOEs should therefore not be used as the sole source of evidence for the position of a glycosidic linkage, *c.q.* for the sequence of glycosyl residues. Using magnitudes of observed interglycosidic NOE effects can lead to wrong conclusions about the primary structure of the carbohydrate. That leaves the detection of long-range $^3J_{CH}$ couplings over glycosidic linkages as the key to sequencing of oligosaccharides by 2-D NMR spectroscopy.

The successful use of long-range $\{^{13}C,^1H\}$ shift-correlation spectroscopy for sequencing an oligosaccharide is dependent on its 1-D ^{13}C spectrum being fully assigned. Given the complete assignment of the ^1H spectrum, the latter task is accomplished by one-bond $\{^{13}C,^1H\}$ shift correlation spectroscopy. The conventional (*i.e.*, ^{13}C-detected) HETCOR experiment [28] yields connectivities between pairs of directly (*via* ^1J) coupled ^{13}C and ^1H nuclei. However, the observed nucleus in HETCOR spectroscopy is ^{13}C, and, therefore, the technique is only applicable if at least 10 to 15 mg of a medium-sized oligosaccharide are available. Recently, the sensitivity problem in the detection of ^1J-connectivities between ^{13}C and ^1H nuclei has been overcome, to a large extent, by the introduction of the so-called *reversed*, *i.e.*, ^1H-detected 2-D $\{^{13}C,^1H\}$ shift-correlation experiment termed HMQC (heteronuclear multiple quantum coherence spectroscopy). The HMQC experiment requires 20 to 30 times less sample than HETCOR to obtain $\{^{13}C,^1H\}$ one-bond shift-correlation maps [29].

The HMQC experiment relies on ^1H detection of multiple- (zero- and double-) quantum coherence created by the basic pulse sequence:

$$\pi/2(^1H) - \Delta - \pi/2(^{13}C) - t_1 - \pi/2(^{13}C) - \text{acquisition}(t_2)(^1H).$$

For small oligosaccharides (with relatively long ^1H T_1's), the dynamic range problem caused by the strong signals of protons not coupled to ^{13}C may be overcome by the suppression of those signals with a bilinear rotation (BIRD) pulse preceding the above pulse sequence [30]; the BIRD pulse inverts the magnetization of protons not coupled to ^{13}C but leaves magnetization of protons coupled to ^{13}C unaffected. For larger oligosaccharides, such as in the cases of the mucin heptasaccharide and the

a.

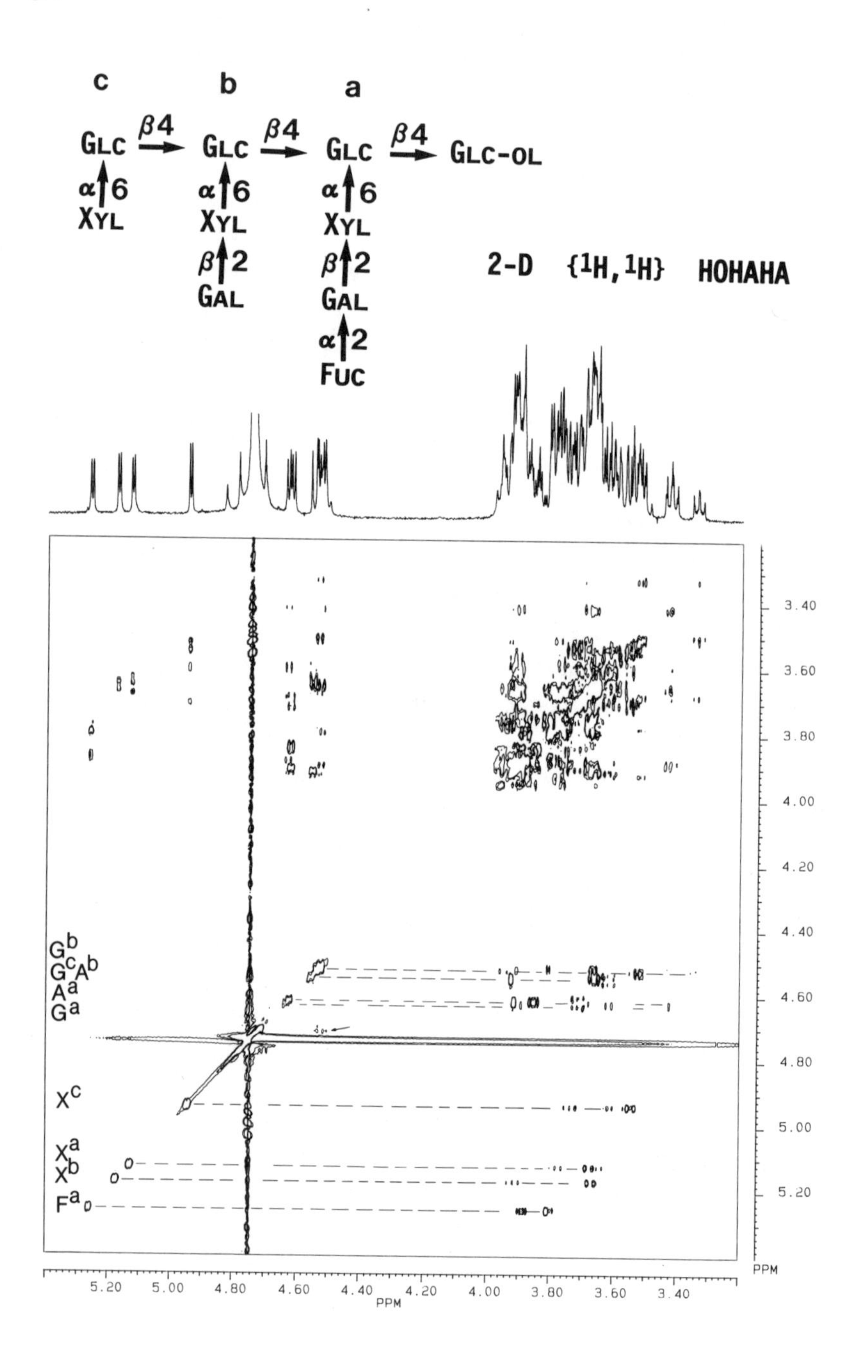
c
b
a
GLC
β4
GLC
β4
GLC
β4
GLC-OL
α6
XYL
β2
GAL
α2
FUC
2-D {1H,1H} HOHAHA
G^b
G^cA^b
A^a
G^a
X^c
X^a
X^b
F^a
3.40
3.60
3.80
4.00
4.20
4.40
4.60
4.80
5.00
5.20
PPM
5.20
5.00
4.80
4.60
4.40
4.20
4.00
3.80
3.60
3.40

xyloglucan decasaccharide, the BIRD pulse was found to be detrimental to the sensitivity of the HMQC experiment because of negative NOE effects during the delay between the BIRD pulse and the HMQC sequence.

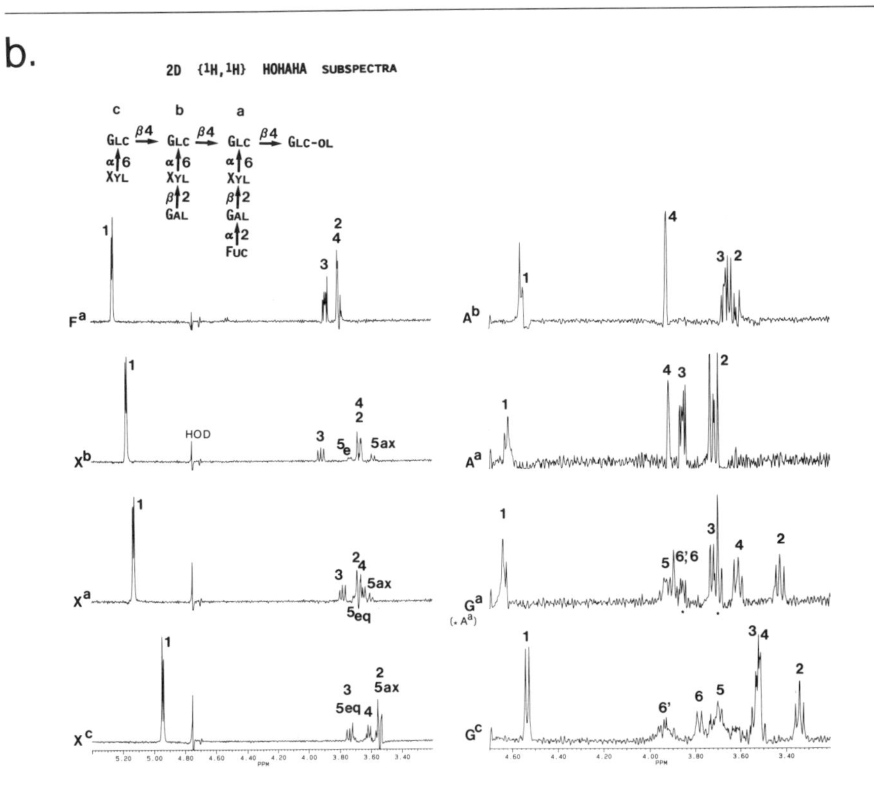

FIGURE 2. a. 2-D HOHAHA spectrum of the decasaccharide isolated from the xyloglucan of rapeseed hulls. The structure of the compound and the conventional 1-D ^{1}H spectrum are shown on top of the contour map. Horizontal, dashed lines connect signals from protons of a particular glycosyl residue. The sample contained 4 mg of the decasaccharide in 0.4 ml D_2O at pD 7 and 27°C. The HOHAHA experiment was performed in the phase-sensitive mode (TPPI) at 500 MHz, on a Bruker AM-500. Data matrix: 256 x 2048; 80 scans per t_1 value. Spin locking was achieved by the MLEV-17 sequence [26]; the mixing time was 200 msec. Sine bell window multiplication was applied in both dimensions before transformation of the data. Total measuring time: 6 h.

b. Cross sections, through the dashed lines, of the 2-D HOHAHA spectrum. The traces represent the 1-D ^{1}H subspectra for the specified glycosyl [fucosyl (F), xylosyl (X^a, X^b, and X^c), glucosyl (G^a and G^c) and galactosyl (A^a and A^b)] residues. The numbers refer to the protons in the glycosyl rings.

a.

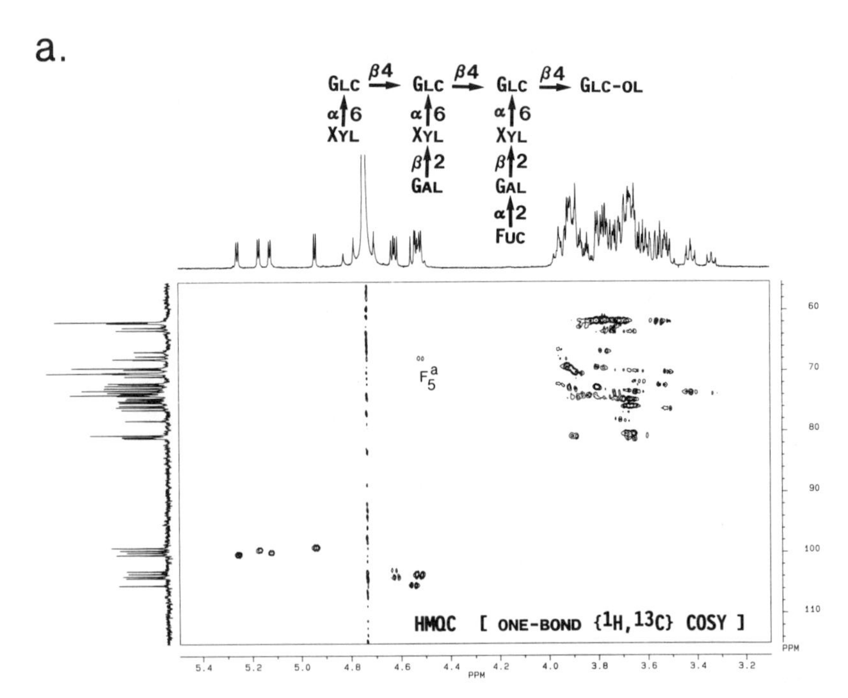
Glc β4 → Glc β4 → Glc β4 → Glc-ol
α↑6 α↑6 α↑6
Xyl Xyl Xyl
β↑2 β↑2
Gal Gal
α↑2
Fuc
F5a
HMQC [ONE-BOND {1H,13C} COSY]
PPM

b.

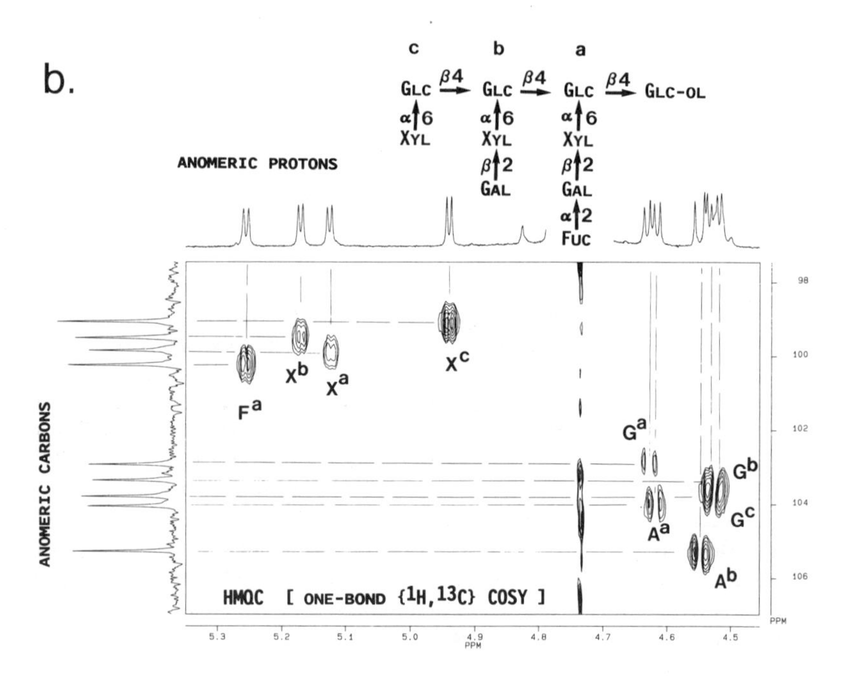
c b a
Glc β4 → Glc β4 → Glc β4 → Glc-ol
α↑6 α↑6 α↑6
Xyl Xyl Xyl
β↑2 β↑2
Gal Gal
α↑2
Fuc
ANOMERIC PROTONS
ANOMERIC CARBONS
Fa
Xb
Xa
Xc
Ga
Gb
Gc
Aa
Ab
HMQC [ONE-BOND {1H,13C} COSY]
PPM

Figure 3.

Residual signals from protons not linked to ^{13}C were removed solely by phase cycling procedures.

The overall HMQC spectrum acquired for the xyloglucan decasaccharide is shown in Fig. 3a. The inset (Fig. 3b) shows the anomeric region of the spectrum revealing the connectivities between the anomeric protons (H1) and anomeric carbons (C1) for each glycosyl residue. The spectrum was recorded in absorption mode, to obtain optimum resolution and sensitivity. To attain the theoretical enhancement in sensitivity over the classical HETCOR experiment, the HMQC experiment was performed with a 5-mm probe with reversed geometry (^{1}H observation coil inside the ^{13}C-tunable decoupling coil). Also, in order to minimize t_1 noise, the sample was not spun during the experiment. The GARP-1 composite-pulse sequence [31] was used for decoupling of ^{13}C during ^{1}H acquisition.
Once both the 1-D ^{1}H and ^{13}C spectra of an oligosaccharide have been completely assigned, long-range $^{3}J_{CH}$ couplings over glycosidic linkages may be used for the sequencing of the oligosaccharide by 2-D NMR spectroscopy. Correlation of ^{13}C and ^{1}H nuclei, scalar-coupled to each other *via* long-range couplings ($^{2}J_{CH}$ and $^{3}J_{CH}$), was conventionally achieved by ^{13}C-detected COLOC (correlation through long-range couplings) spectroscopy [32]. The COLOC pulse sequence discriminates between ^{13}C-^{1}H connectivities through $^{1}J_{CH}$ and long-range J_{CH} couplings by a low-pass J-filter. However, the COLOC experiment requires at least an order ofmagnitude more sample than conventional HETCOR. Even then, reduction of intensity or even cancellation of cross-peaks may occur as a function of $^{1}J_{CH}$.

The ^{1}H-detected heteronuclear multiple-bond connectivity (HMBC)

FIGURE 3. a. HMQC spectrum of the decasaccharide isolated from the xyloglucan of rapeseed hulls. The conventional 1-D ^{1}H and ^{13}C spectra are shown along the horizontal and vertical axes, respectively. The sample was the same as described in the legend to Fig. 2. Data matrix: 128 x 2048; 128 scans per t_1 value. A 24-Hz Gaussian broadening was used in the t_1 dimension; a squared sine bell window shifted over $\pi/3$ was applied in the t_2 dimension. Total measuring time: 5 h.

b. Anomeric region of the HMQC spectrum. Dashed lines indicate the one-bond connectivities between the H1 and C1 atoms. Symbols for residues are as in Fig. 2.

a.

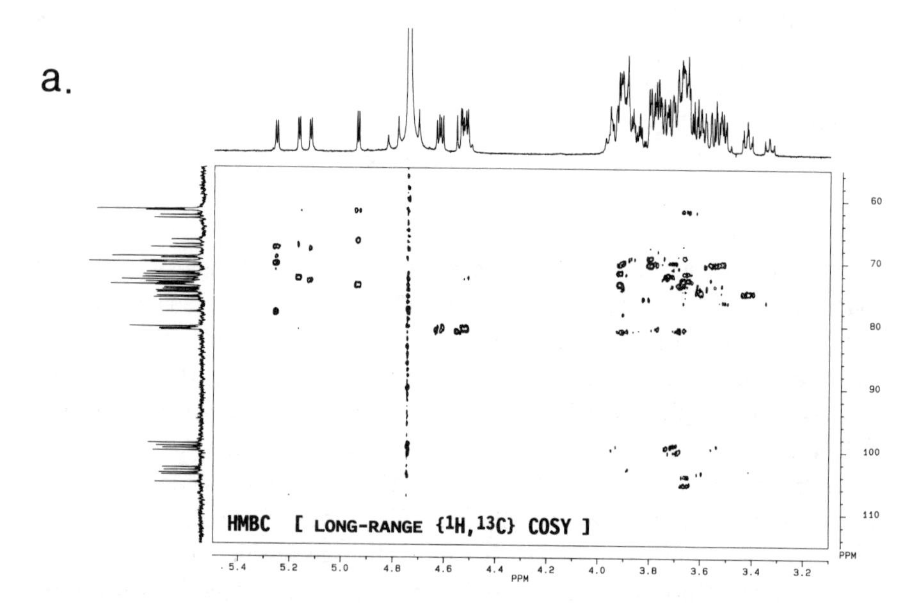
HMBC [LONG-RANGE {1H,13C} COSY]
PPM

b.

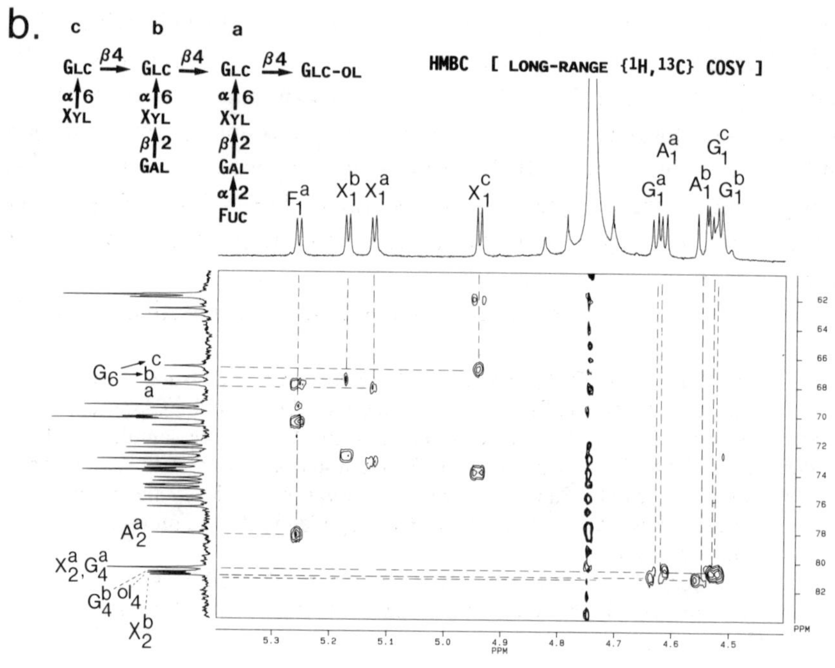
c
b
a
GLC
β4
GLC
β4
GLC
β4
GLC-OL
α6
XYL
β2
GAL
α2
FUC
HMBC [LONG-RANGE {1H,13C} COSY]
PPM

Figure 4.

experiment is 50 to 100 times more sensitive than the COLOC experiment [33,34]. The HMBC pulse sequence

$$\pi/2(^1H) - \Delta_1 - \pi/2(^{13}C) - \Delta_2 - \pi/2(^{13}C) - t_1/2 - \pi(^1H) - t_1/2 - \pi/2(^{13}C) - acquisition(t_2)(^1H)$$

generates multiple-quantum coherence between ^{13}C and ^1H coupled through $^2J_{CH}$ and $^3J_{CH}$, which is detected by ^1H observation. Spectrometer stability is of the utmost importance for this experiment, since unwanted signals can only be minimized by phase-cycling and subtraction of alternate scans. The sequencing of the xyloglucan decasaccharide by HMBC, including the determination of the branch location of the Fucα(1→2)Galβ(1→2) moiety, is illustrated in Fig. 4. The sequencing, by HMBC, of the salivary mucin heptasaccharide is illustrated in Fig. 5. The connectivities between the anomeric protons and the carbons at the sites of substitution provide most of the sequence information. However, in the case of the mucin heptasaccharide, the NeuAc residue (a ketose) does not have an anomeric proton. Nevertheless, the linkage between NeuAc and Gal^3 is recognized to be (2→3) from the long-range connectivity between C2 (anomeric carbon) of NeuAc and H3 of Gal^3 at the site of substitution (see Fig. 5a).

In summary, the HMQC experiment, providing a one-bond $\{^{13}C,^1H\}$ correlation map, is generally sufficient to unambiguously assign all resonances in the ^{13}C-NMR spectrum, provided that the entire ^1H spectrum was assigned by the COSY/HOHAHA approach. The tracing of long-range interglycosidic $^3J_{CH}$ from the HMBC spectrum completes the primary structural analysis by providing the linkage positions and the sequence of the glycosyl residues in the complex carbohydrate.

FIGURE 4. a. HMBC spectrum of the decasaccharide isolated from the xyloglucan of rapeseed hulls. The conventional 1-D ^1H and ^{13}C spectra are shown along the horizontal and vertical axes, respectively. The ^1H spectrum had been assigned by HOHAHA (see Fig. 2), the ^{13}C spectrum subsequently by HMQC (see Fig. 3). The sample was the same as described in the legend to Fig. 2. Data matrix: 128 x 2048; 176 scans per t_1 value. A 12-Hz Gaussian broadening was used in the t_1 dimension; a squared sine bell window shifted over $\pi/6$ was applied in the t_2 dimension. Total measuring time: 16 h.

b. Anomeric region of the HMBC spectrum. Connectivities arising from $^3J_{CH}$ couplings across the glycosidic bonds are indicated. Symbols for residues are as in Fig. 2.

a.

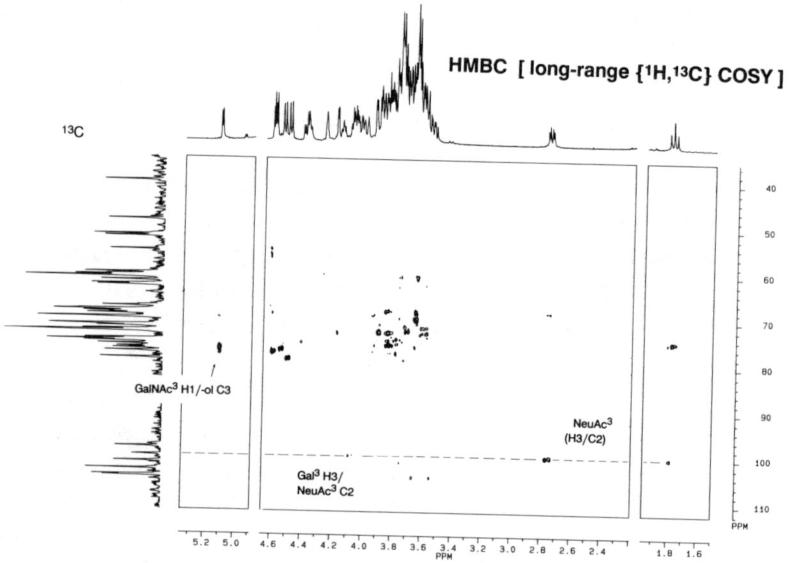

HMBC [long-range {^{1}H,^{13}C} COSY]

b.

HMBC [long-range {^{1}H,^{13}C} COSY]

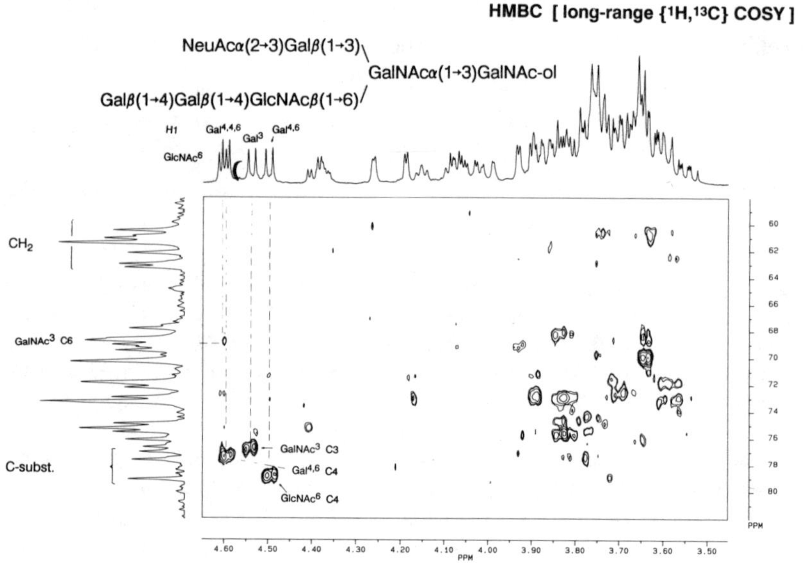

NeuAcα(2→3)Galβ(1→3)\
GalNAcα(1→3)GalNAc-ol
Galβ(1→4)Galβ(1→4)GlcNAcβ(1→6)/

Figure 5.

Locating naturally occurring O-acyl groups in complex carbohydrates —— Finally we illustrate the usefulness of the combined COSY/HOHAHA and HMQC/HMBC approach for locating the sites of naturally occurring *O*-acyl groups in a complex carbohydrate, *e.g.*, in a highly esterified trisaccharide isolated from the stem bark of *Mezzettia leptopoda (Annonaceae)* [35].

C **B**

{3,4-di-OAc-L-Rha*p*}-α(1→3)-{2,4-di-OAc-L-Rha*p*}-α(1→3)-

{4-OHex-L-Rha*p*}-α(1→octyl)

A

The ^{1}H spectrum of the trirhamnoside was recorded in C_6D_6/CD_3OD (4:1), and completely assigned by COSY and HOHAHA as described above (results not shown). The HMQC experiment then yielded the complete assignment of the protonated carbons in the ^{13}C spectrum. Fig. 6 shows two portions of the HMBC spectrum recorded for the trirhamnoside. Fig. 6a shows, *i.a.*, the cross peaks between the anomeric protons and the carbons of the adjacent rhamnosyl residues at the site of substitution (H1C-C3B and H1B-C3A, respectively), as well as the connectivity between H1A and Ca assigned to the methylene group of the octyl aglycon. This establishes both interglycosidic linkages in the trisaccharide as (1→3) and completes the assignment of the ^{1}H and ^{13}C spectra. Fig. 6b shows the portion of the HMBC spectrum where connectivities between ester carbonyl carbons and rhamnosyl ring protons are observed. The long-chain acyl (hexanoyl) group has its carbonyl resonance at 173.5 ppm, while the four acetyl groups in the molecule all resonate at $\delta \approx 170.7$ ppm. The acetyls were placed at C2 and

---

FIGURE 5. a. HMBC spectrum of the heptasaccharide isolated from the salivary mucins of the Chines swiftlet. The conventional 1-D ^{1}H and ^{13}C spectra are shown along the horizontal and vertical axes, respectively. The ^{1}H spectrum had been assigned by COSY (see Fig. 1) and HOHAHA, the ^{13}C spectrum subsequently by HMQC. The sample was the same as described in the legend to Fig. 1. Data matrix: 128 x 2048; 128 scans per t_1 value. A 12-Hz Gaussian broadening was used in the t_1 dimension; a squared sine bell window shifted over $\pi/3$ was applied in the t_2 dimension. Total measuring time: 16 h.

b. Anomeric region of the HMBC spectrum. Connectivities arising from $^3J_{CH}$ couplings across the glycosidic bonds are indicated.

a.

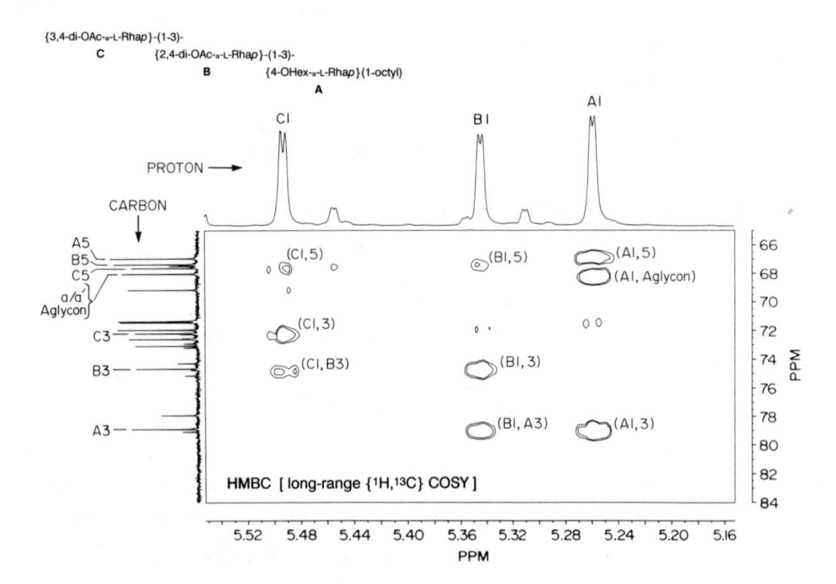
{3,4-di-OAc-α-L-Rhap}-(1-3)-
C
{2,4-di-OAc-α-L-Rhap}-(1-3)-
B
{4-OHex-α-L-Rhap}(1-octyl)
A
PROTON
CARBON
C1
B1
A1
A5
B5
C5
a/a'
Aglycon
C3
B3
A3
(C1,5)
(B1,5)
(A1,5)
(A1, Aglycon)
(C1,3)
(C1,B3)
(B1,3)
(B1,A3)
(A1,3)
HMBC [long-range {1H,13C} COSY]
PPM

b.

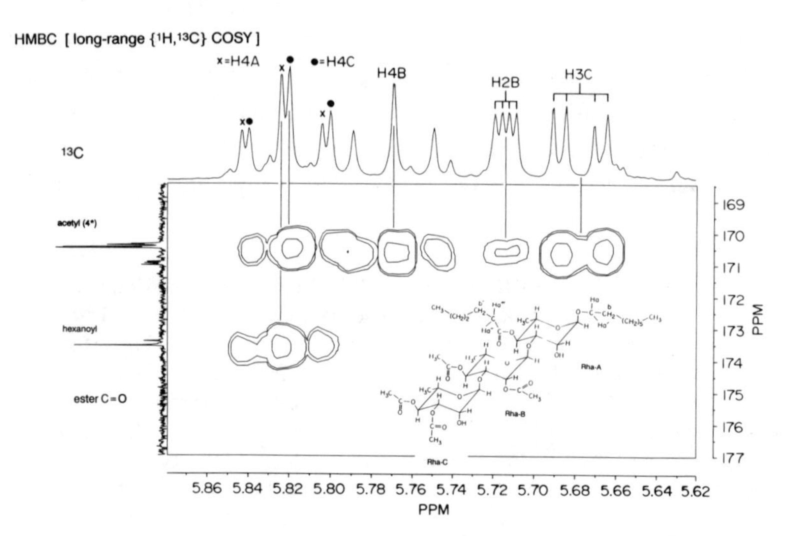
HMBC [long-range {1H,13C} COSY]
x=H4A
●=H4C
H4B
H2B
H3C
13C
acetyl (4")
hexanoyl
ester C=O
Rha-A
Rha-B
Rha-C
PPM

Figure 6.

C4 of Rha-B, and at C3 and C4 of Rha-C, while the hexanoyl group was found to be attached to C4 of Rha-A. Thus, tracing the long-range connectivity between an ester carbonyl carbon and a proton in the glycosyl ring reveals the site at which the ester group is located.

CONCLUSIONS

NMR spectroscopy is eminently suited for providing both primary and secondary structural information on complex carbohydrates in solution. Complete resonance assignments and primary structure determination are a prerequisite for the analysis of the solution conformation based on quantitation of NOEs.

The combination of multiple-pulse, ^{1}H- and ^{13}C-NMR spectroscopic techniques allows the *de-novo* sequencing of a carbohydrate as large as a decasaccharide (*c.q.*, a polysaccharide with a repeat unit as large as decasaccharide), provided that a few μmol of pure substance are available for the analysis. 2-D HOHAHA spectroscopy, along with double-quantum filtered {^{1}H,^{1}H} COSY, are well suited to yield the complete assignment of the ^{1}H-NMR spectrum of a carbohydrate; the HOHAHA technique permits subspectral editing for each constituting monosaccharide. Once all the multiplet patterns in the ^{1}H-NMR spectrum have been deciphered, the identification of each of the *J*-coupled spin-networks in terms of the constituent glycosyl residues is possible, along with their anomeric configurations.

FIGURE 6. a. Anomeric region of the HMBC spectrum of the trirhamnoside isolated from the stem bark of *Mezzettia leptopoda* (the sample for NMR analysis contained 20 mg in 0.4 ml C_6D_6/CD_3OD, 4:1, *v/v*, at 25°C). The conventional 1-D ^{1}H and ^{13}C spectra are shown along the horizontal and vertical axes, respectively. The ^{1}H spectrum had been assigned by HOHAHA, the ^{13}C spectrum subsequently by HMQC. Data matrix: 200 x 2048; 192 scans per t_1 value. A 100-Hz Gaussian broadening was used in the t_1 dimension; a non-shifted sine bell window was applied in the t_2 dimension. Total measuring time: 16 h.

b. Ester carbonyl region of the HMBC spectrum. Peaks arising from $^{3}J_{CH}$ couplings between C=O and protons indicate the positions of the acyl groups in the rings.

Reversed $\{^{13}C,^{1}H\}$ correlation-spectroscopy is a highly sensitive method that provides $\{^{13}C,^{1}H\}$ chemical-shift correlation maps of an oligo saccharide through the ^{1}H-detection of heteronuclear multiple quantum coherences (HMQC and HMBC). Once the HOHAHA and HMQC deduced assignments of the ^{1}H and ^{13}C spectra of a carbohydrate are complete, the interglycosidic long-range heteronuclear couplings can then be used to link the glycosyl residues together. The HMBC spectrum permits the deduction of the sequence of the glycosyl residues, and also the location of most types of appended groups.

ACKNOWLEDGMENTS

The author is indebted to Mr. William York and Mr. Donald Powell for recording the NMR spectra of the xyloglucan oligosaccharide and the *Mezzettia leptopoda* trisaccharide, respectively, and to Mr. Dennis Warrenfeltz for continuous technical support. Thanks are due to Mrs. Carol Gubbins-Hahn for preparing the line-drawings and figures, and to Dr. Anne-Marie Strang for critical review of the manuscript.

REFERENCES

1. a) Knight P (1989) *Biotechnol* **7**: 35-40. b) Feizi T (1985). *Nature* **314**: 53-57. c) Feizi T, Childs RA (1985). *Trends Biochem Sci* **10**: 24-29.
2. Rademacher TW, Parekh RB, Dwek RA (1988). *Ann Rev Biochem* **57**: 785-838.
3. Darvill AG, Albersheim P (1984). *Ann Rev Plant Physiol* **35**: 243-275.
4. Bock K, Thøgersen H (1982). *Ann Rep NMR Spectrosc* **13**: 1-57.
5. Bradbury JH, Jenkins GA (1984). *Carbohydr Res* **126**: 125-156.
6. Carver JP, Grey AA (1981). *Biochemistry* **20**: 6607-6616.
7. Vliegenthart JFG, Dorland L, Van Halbeek H (1983). *Adv Carbohydr Chem Biochem* **41**: 209-374.
8. Van Halbeek H (1984). *Biochem Soc Trans* **12**: 601-605.
9. Argade S, Hopfer RL, Strang AM, Van Halbeek H, Alhadeff JA (1988). *Arch Biochem Biophys* **266**: 227-247.
10. Van Halbeek H, Breg J, Vliegenthart JFG, Klein A, Lamblin G, Roussel P (1988). *Eur J Biochem* **177**: 443-460.
11. Green ED, Adelt G, Baenziger JU, Wilson S, Van Halbeek H (1988). *J Biol Chem* **263**: 18253-18268.
12. a) Yu RK, Koerner TAW, Scarsdale JN, Prestegard JH (1986). *Chem Phys Lipids* **42**: 27-48. b) Koerner TAW, Prestegard JH, Yu RK (1987). *Methods Enzymol* **138**: 38-59.
13. Homans SW, Dwek RA, Rademacher TW (1987). *Biochemistry* **26**: 6571-6578.

14. Dabrowski J (1987). In Croasmun WR, Carlson RMK (eds): "Two-dimensional NMR spectroscopy — applications for chemists and biochemists," New York: VCH, p 349-386.
15. Bush CA (1988). *Bull Magn Reson* **10**: 73-95.
16. Byrd RA, Egan W, Summers MF, Bax A (1987). *Carbohydr Res* **166**: 47-58.
17. Wieruszeski JM, Michalski JC, Montreuil J, Strecker G, Peter-Katalinic J, Egge H, Van Halbeek H, Mutsaers JHGM, Vliegenthart JFG (1987). *J Biol Chem* **262**: 6650-6657.
18. a) Wagner G (1983). *J Magn Reson* **55**: 151-156. b) King G, Wright PE (1983). *J Magn Reson* **54**: 328-332.
19. Homans SW, Dwek RA, Fernandes DL, Rademacher TW (1984). *Proc Natl Acad Sci USA* **81**: 6286-6289.
20. Braunschweiler L, Ernst RR (1983). *J Magn Reson* **53**: 521-528.
21. Bax A, Davis DG (1985). *J Magn Reson* **65**: 355-360.
22. York WS, Oates JE, Van Halbeek H, Darvill AG, Albersheim P, Tiller PR, Dell A (1988). *Carbohydr Res* **173**: 113-132.
23. York WS, Van Halbeek H, Warrenfeltz DL, Darvill AG, Albersheim P (1989). (submitted for publication).
24. Subramanian S, Bax A (1987). *J Magn Reson* **71**: 325-330.
25. Homans SW, Dwek RA, Boyd J, Soffe N, Rademacher TW (1987). *Proc Natl Acad Sci USA* **84**: 1202-1205.
26. Davis DG, Bax A (1985). *J Am Chem Soc* **107**: 7197-7198
27. Bush CA, Yan ZY, Rao BNN (1986). *J Am Chem Soc* **108**: 6168-6173.
28. Bax A (1982). "Two-dimensional nuclear magnetic resonance in liquids." Delft, Delft Univ Press, p 51-68.
29. Bax A, Subramanian S (1986). *J Magn Reson* **67**: 565-569.
30. Lerner L, Bax A (1987). *Carbohydr Res* **166**: 35-46.
31. Shaka AJ, Keeler J (1987). *Prog NMR Spectrosc* **19**: 47-129.
32. Kessler H, Griesinger C, Zarbock J, Loosli HR (1984). *J Magn Reson* **57**: 331-336.
33. Bax A, Summers MF (1986). *J Am Chem Soc* **108**: 2093-2094.
34. Summers MF, Marzilli LG, Bax A (1986). *J Am Chem Soc* **108**: 4285-4294.
35. Powell DAP, York WS, Etse JT, Waterman PG, Van Halbeek H (1989). (submitted for publication).

Frontiers of NMR in Molecular Biology, pages 215-224

STUDIES OF THE 5S RNA FROM *ESCHERICHIA COLI* BY NMR[1]

Peter B. Moore

Departments of Chemistry, and Molecular Biophysics and Biochemistry, Yale University, New Haven, CT 06511.

ABSTRACT. Imino proton NMR spectroscopy shows that the secondary structure of the helix V/loop E region of the 5S RNA from *E. coli* is structured, but is not base paired in the manner that current, phylogenetically derived models indicate it should be. Recent results with RNAs made using T7 RNA polymerase suggest that the enzyme makes errors at a spectroscopically significant rate when functioning *in vitro*.

INTRODUCTION

Recent developments make it likely that our understanding of RNA will increase dramatically in the near future. First, everyone's level of consciousness has been raised by the discovery of RNA catalysis (1; 2), and enzymatic properties have been demonstrated recently in RNAs that are remarkably small (3). Second, the number of targets for structure determination has been further expanded by the discovery of a large number of small, stable RNAs that participate in gene expression in eukaryotes (see 4). Third, the "availability barrier" has been breached. RNA polymera-

[1]This work was supported by grants from NSF (DMB-860823) and NIH (AI-09167, GM-22778)

ses from bacteriophages such as T7 can be used to prepare 10's of milligrams of RNAs of almost any sequence in vitro (5; see below), and chemical synthesis of RNA should soon become as routine as DNA synthesis has been for several years (e.g. 6).

The last decade has also seen tremendous advances in the methodology for the determination of the structures of biological macromolecules by NMR, as the papers in this volume attest. It is inevitable that NMR will become an increasingly important source of information about RNA structure as time goes by. This paper summarizes our recent studies of the 5S RNA system from E. coli by NMR.

MATERIALS AND METHODS

RNAs.

The methods used for producing 5S RNA, 5S RNA fragment 1 and ^{15}N-labelled fragment 1 have been described previously (7, 8), as have techniques for making RNAs using T7 RNA polymerase (9).

Spectroscopic samples and methods.

Samples were prepared for spectroscopy by dialysis into 0.1 M KCl, 4mM $MgCl_2$, 5mM cacodylate, pH 7.0. RNA concentrations of 1mM or higher were obtained by ultrafiltration following dialysis. Spectra were obtained on the 490 MHz and 500 MHz spectrometers of the Yale Chemical Instrumentation Center operating in the Fourier transform mode. One-dimensional imino proton spectra were obtained using the twin pulse method (10).

RESULTS

Imino proton spectroscopy.

The downfield, imino proton region (10 - 15 ppm) is by far the easiest part of a nucleic acid's spectrum to resolve. Because only guanosine and uridine residues have imino protons, there is about half an imino protons per base in a typical RNA, and their resonances are dispersed over 5 ppm. By way of contrast, there are 5 non-anomeric sugar protons per residue in RNA, and their resonances are found between 3.6 and 4.9 ppm.

A second attraction of imino proton spectroscopy is its connection with nucleic acid secondary structure. Imino protons have to be protected from solvent exchange in order to be observable by NMR, and most obtain that protection by participation in hydrogen-bonded base pairs. The imino proton spectrum of an RNA is thus an expression of its secondary structure. For both reasons most investigations of a nucleic acids by NMR start with a study of its imino proton spectrum.

Our work has concentrated on an RNA derived from the 5S RNA of E. coli by limited nucleolytic digestion, or more recently, by genetic engineering techniques (9). This 60 base RNA constitutes half the parent molecule, and has a structure that is similar to that of the same sequences in the parent molecule (Figure 1) (7,10). Figure 2 compares

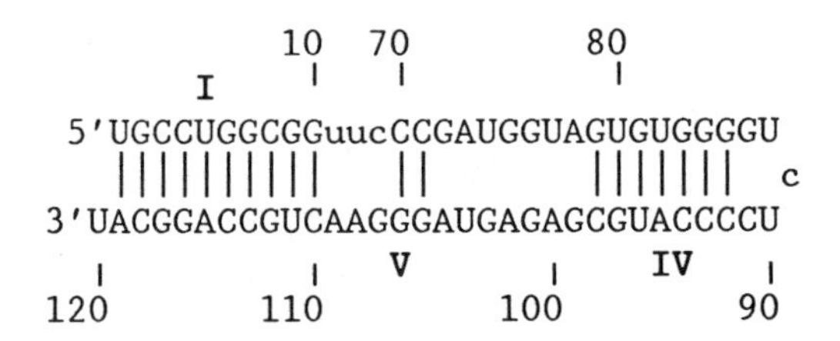

Figure 1. Sequences of fragment 1 and pDG07 RNA. The sequence of pDG07 RNA is shown with its bases numbered according to the sequence of intact 5S RNA. Helical regions are designated by roman numerals. [Loop E lies between helices IV and V.] Lower case bases are present in pDG07 RNA, but absent from fragment 1.

the imino proton spectra of fragment 1, the nucleolytic product, and that of pDG07 RNA, the genetically engineered molecule.

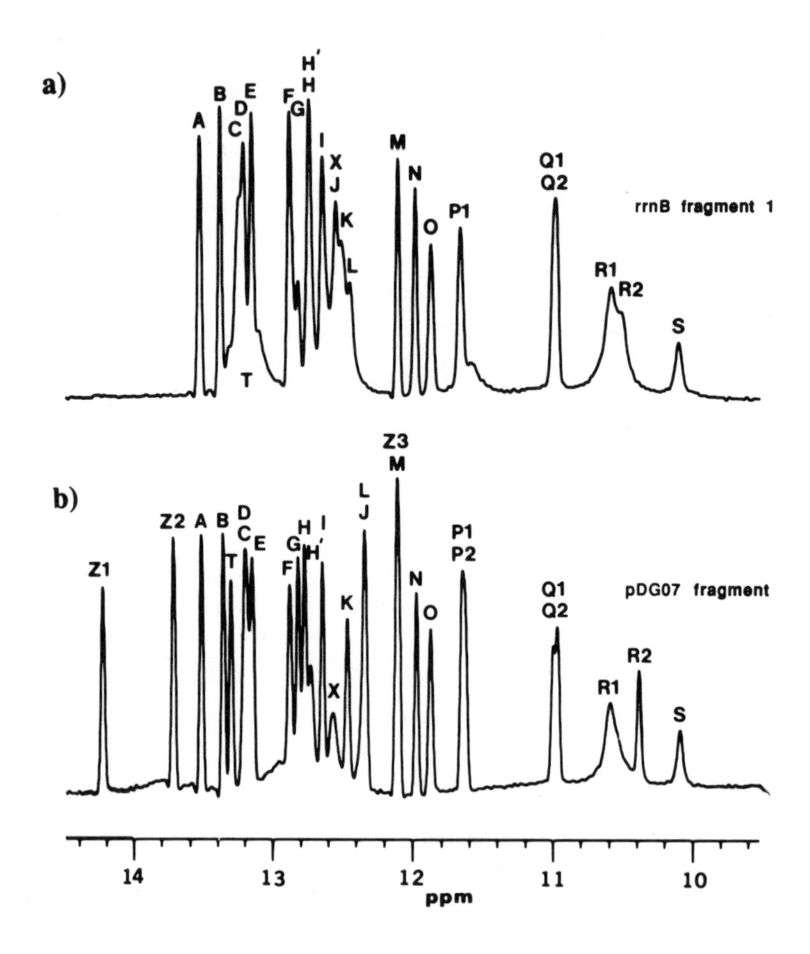

Figure 2. Comparison of the downfield spectra of fragment 1 and pDG07 RNA. Spectrum a is the downfield spectrum of normal fragment 1 at 303 K. Its resonances are identified by letters following the convention of Kime and Moore (7). Spectrum b is that of pDG07 at 303 K. Resonances in the two spectra having the same name arise from homologous protons. Resonances Z1, Z2, and Z3 represent the base pairs created by the 3 base insert used to link base 10 with base 70 in the normal fragment sequence (see Figure 2). (Reproduced from (11) with permission of the copyright holder, the American Chemical Society.

Initial Assignments.

About 1980, Redfield and his colleagues demonstrated that NOE connectivities between imino proton resonances could be used to assign imino proton spectra (see 12). We applied Redfield's method to the spectrum of fragment 1 as soon as we obtained it, and the resonances that belong to its helix I and helix IV portions (see Figure 1) were rapidly identified (7, 13, 14).

Left unassigned were roughly half the resonances in the spectrum, many of which originate in the region between helices I and IV, i.e. in helix V/loop E. Inspection of Figure 1 will reveal that only a few "standard" base pairs are possible in this part of the molecule, and while there is evidence that its two strands are paired, the structure formed is not conventional double helix (15). Its irregularity results in a lack of NOE connectivities between resonances that makes assignment problematical. Helix V/loop E is arguably the part of fragment 1 from which one is most likely to learn interesting lessons about RNA, but it is in regions such as this that assignment techniques are at their weakest.

Assignments in helix V/loop E.

Penghua Zhang has recently completed the assignment of the resonances that originate in the helix V/loop E region of fragment 1. The data that have done the most to settle the issue come from experiments with ^{15}N-labelled samples. Imino proton resonances, which are singlets in ordinary nucleic acids, become 85 Hz doublets in ^{15}N-labelled material due to coupling. Several techniques exist for determining the correlations between ^{15}N chemical shifts and the chemical shifts of the protons that are coupled to them (16). These correlations are interesting because the chemical shifts of the N1's of guanines and the N3's of uridines are cleanly separated in the ^{15}N frequency domain (17). Thus determination of the ^{15}N chemical shift of the N coupled to a given imino proton identifies it as to base type.

Zhang's experiments involved the examination of samples partially labelled with ^{15}N. At low temperatures in the presence of Ca^{2+} she detected some "new" resonances and "new" imino-imino NOEs. This information, supplemented by spectra taken on RNAs with point mutations in the helix V/loop E region, and the information obtained earlier using

[15]N-labelled samples (8, 18, 19) finished the job (20). The results are summarized in Figure 3.

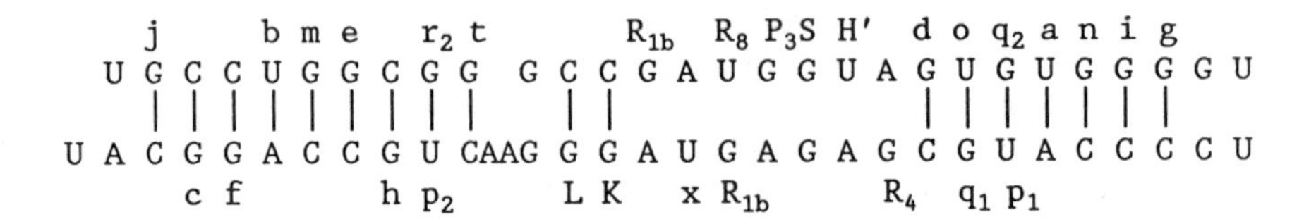

Figure 3) Assignments of Resonances in Fragment 1. The sequence of fragment 1 is shown with its imino proton assignments indicated. The assignments in capital letters are those established by Zhang (20). Resonances not identified in Figure 2 are seen only at temperatures below 283 K.

On the production of RNA using T7 RNA polymerase.

T7 polymerase is a robust enzyme, which transcribes any DNA sequence downstream of a T7 promoter, and ignores all others. The enzyme can be over produced (21), and a number of methods have been developed for using it to make RNA both in vivo (22) and in vitro (5) in biophysically useful amounts.

Like many others we have been using T7 RNA polymerase to make RNAs for both NMR and crystallographic study, but have been unimpressed by the quality of the material produced (9). [The quantity of the product has been as impressive as advertised!] The linewidths of the imino proton NMR resonances of T7 products are invariably much broader than observed for comparable naturally produced RNAs, and the larger the T7 transcript, the worse it seems to be (9).

T7 RNA polymerase reaction mixtures contain large amounts of abortive initiation products in addition to the desired full length transcripts. Could RNA impurities be responsible for the linewidths we see? Dan Gewirth purified some of his transcripts on Sephadex columns in urea at elevated temperature. The linewidths in the spectra of the product were unaltered. Susan White recently purified one of Gewirth's T7 transcripts on denaturing gels. The linewidths of the spectrum her material gave are similar to the linewidths Gewirth observed earlier (9).

These findings suggest that T7 RNA polymerase may be error-prone under the conditions used for transcription *in vitro*, and a low level of sequence variation might account for the linewidths we observe. This reasoning led us to question whether T7 RNA polymerase was capable of making high quality transcripts under any conditions. To examine this issue the truncated 5S RNA gene in pDG07 was inserted into pAR3056 (23) downstream of its T7 promoter. This construct was transduced into *E. coli* BL21(DE3) (22), which carries a gene for T7 RNA polymerase in its chromosome under lac control. Induction of the T7 RNA polymerase gene by IPTG resulted in the production of the RNA desired.

Table 1 compares the linewidths of a number of resonances in the downfield spectrum of pDG07 RNA made by the polymerase of *E. coli in vivo*, T7 RNA polymerase *in vivo*, and T7 polymerase *in vitro*. The *in vivo* T7 product gives a spectrum comparable to that of "normal" pDG07 RNA in every way. This finding supports the hypothesis that the conditions we use for T7 transcription *in vitro*, which were optimized for the amount of RNA produced, make the transcription process error-prone. The fact that T7 RNA polymerase transcription *in vivo* is accurate indicates that the intracellular environmental provides either the ionic conditions the enzyme needs or some host factors that ensure faithful transcription, and gives us hope that a "cure" can be found for the problem we have been having.

DISCUSSION

Our experience with fragment 1 shows that NOE methods, by themselves, are unlikely to assign imino proton resonances originating in regions that do not have conventional double helical structures. Even if the bases in such regions give detectable resonances, NOE connectivities are likely to be too fragmented, and may disappear altogether. We would not have succeeded had ^{15}N labelling not been used, and we were significantly assisted by our ability to create specific 5S RNA mutants.

TABLE 1
LINE WIDTHS OF SINGLE PROTON RESONANCES IN pDG07-LIKE RNAS PRODUCED USING T7 RNA POLYMERASE IN VIVO AND IN VITRO[a]

Resonance	RNAs tested			
			in vitro	
	standard	in vivo	gel	column
Z1	18.5	21.5	31.6	41.7
Z2	19.0	18.1	34.6	37.7
A	18.1	17.8	32.0	32.8
N	19.8	18.5	-	-
O	20.6	19.8	-	-

[a]The full widths of resolved resonances that correspond to single protons were measured (in Hz) at half height. The spectra of the 4 samples in question were taken at the same temperature and approximately the same RNA concentration. The standard is pDG07 RNA produced in vivo by E. coli enzymes (11). The in vivo RNA is pDG07 RNA produced in vivo by T7 RNA polymerase (see text). The in vitro products are pJK1 RNA (9), which has a sequence closely similar to that of pDG07 RNA. One sample was purified on gels, and the other on a Sephadex column in urea. Entries marked "-" were not evaluated because the resonances in question were not well enough resolved from their neighbors.

Our experience with the non-exchangeable protons in fragment 1, which is not discussed above, has further emphasized the pernicious connection in nucleic acid spectroscopy between one's expectations about a molecule's structure and the assignment of its spectrum. The non-exchangeable proton spectra of nucleic acids with novel, unexpected structures probably cannot be assigned without resort to isotopic labelling. What is needed at this point is the development of a systematic, economical approach to the assignment problem that will not be defeated by structural irregularities.

In our view, the hypothesis that T7 RNA polymerase is error-prone is strengthened by our recent results. But there is clearly hope these problems can be resolved; the enzyme works correctly in vivo. Perhaps with some modest alterations of conditions in vitro the enzyme will fulfill its promise as an economical means for making RNA samples for biophysical investigation.

REFERENCES

1. Kruger, K., Grabowski, P.J., Zaug, A.J., Sands, J., Gottschling, D.E. & Cech, T.R. (1982). Self splicing RNA: autoexcision and autocyclization of the ribosomal RNA intervening sequence of Tetrahymena. Cell 31: 147.
2. Guerrier-Takada, C., Gardiner, K., Marsh, T., Pace, N. & Altman, S. (1983). The RNA moiety of ribonuclease P is the catalytically active subunit of the enzyme. Cell 35: 849.
3. Uhlenbeck, O.C. (1987). A small catalytic oligoribonucleotide. Nature 328: 596.
4. Padgett, R.A., Grabowski, P.J., Konarska, M.M., Seiler, S. & Sharp, P.A. (1986). Splicing of messenger RNA precursors. Ann Rev Biochem 55: 1119.
5. Milligan, J.F., Groebe, D.R., Wetherell, G.W. & Uhlenbeck, O.C. (1987). Oligoribosnucleotide synthesis using T7 RNAS polymerase and synthetic DNA templates. Nucleic Acids Res 15: 8783.
6. Usman, N., Ogilvie, K.K., Jiang, M.-Y. & Cedergren, R.J. (1987). Automated chemical synthesis of long oligoribonucleotides using 2′-O-silylated ribonucleoside 3′-O′phosphoramidites on a controlled-pore glass support: synthesis of a 43-nucleotide sequence similar to the 3′-half molecule of an Escherichia coli formylmethionyl tRNA. J Am Chem Soc 109: 7845.
7. Kime, M.J. & Moore, P.B. (1983). Nuclear Overhauser experiments at 500MHz on the downfield proton spectrum of a ribonuclease-resistant fragment of 5S ribonucleic acid. Biochemistry 22: 2615.
8. Gewirth, D.T., Abo, S.R., Leontis, N.B. & Moore, P.B. (1987). The secondary structure of 5S RNA: NMR experiments on RNA molecules partially labelled with 15N. Biochemistry 26: 5213.
9. Gewirth, D.T. & Moore, P.B. (1988). Exploration of the L18 binding site on 5S RNA by deletion mutagenesis. Nucl Acids Res 16: 10717.
10. Kime, M.J. & Moore, P.B. (1983). Physical evidence for a domain structure in Escherichia coli 5S RNA. FEBS Letters 153: 199.
11. Gewirth, D.T. & Moore, P.B. (1987). Effects of mutation

on the downfield proton nuclear magnetic resonance spectrum of the 5S RNA of Escherichia coli. Biochemistry 26: 5657.

12. Patel, D.J., Shapiro, L. & Hare, D. (1987). DNA and RNA: NMR studies of conformation and dynamics in solution. Quart Rev Biophys 20: 36.
13. Kime, M.J. & Moore, P.B. (1983). Nuclear Overhauser experiments at 500MHz on the downfield proton spectrum of 5S ribonucleic acid and its complex with ribosomal protein L25. Biochemistry 22: 2622.
14. Kime, M.J. & Moore, P.B. (1984). Escherichia coli ribosomal 5S RNA-protein L25 nucleoprotein complex: effects of RNA binding on the protein structure and the nature of the interaction. Biochemistry 23: 1688.
15. Garrett, R.A., Douthwaite, S. & Noller, H.F. (1981). Structure and role of 5S RNA-protein complexes in protein biosynthesis. Trends Biochem Sci 5: 137.
16. Griffey, R.H., Redfield, A.G., Loomis, R.E. & Dahlquist, F.W. (1985). Nuclear Magnetic Resonance Observation and dynamics of specific amide protons in T4 lysozyme. Biochemistry 24: 817.
17. Gonnella, N.C., Birdseye, T.R., Nee, M. & Roberts, J.D. (1982). ^{15}N NMR study of a mixture of uniformly labeled tRNAs. Proc Nat Acad Sci USA 79: 4834.
18. Kime, M.J. (1984). Assignment of resonances in the E. coli 5S RNA fragment proton NMR spectrum using uniform nitrogen-15 enrichment. FEBS Letters 173: 342.
19. Kime, M.J. (1984). Assignment of resonances of exchangeable protons in the NMR spectrum of the complex formed by E. coli ribosomal protein L25 and uniformly nitrogen-15 enriched 5S RNA fragment. FEBS Letters 175: 259.
20. Zhang, P. & Moore, P.B. (1989). An NMR study of the helix V-loop E region of the 5S RNA from E. coli. manuscript in preparation.
21. Davanloo, P., Rosenberg, A.H., Dunn, J.J. & Studier, F.W. (1984). Cloning and expression of the gene for bacteriophage T7 RNA polymerase. Proc Nat Acad Sci USA 81: 2035.
22. Studier, F.W. & Moffat, B.A. (1986). Use of bacteriophage T7 RNA polymerase to direct selective high-level expression of cloned genes. J Mol Biol 189: 113.
23. Steen, R., Jemiolo, D.K., Skinner, R.H., Dunn, J.J. & Dahlberg, A.E. (1986). Expression of plasmid-coded mutant ribosomal RNA in E. coli: choice of plasmid vectors and gene expression systems. Prog. Nucleic Acid Res Mol Biol 33: 1.

Frontiers of NMR in Molecular Biology, pages 225-238

APPLICATIONS of IMINO PROTON EXCHANGE to NUCLEIC ACID KINETICS and STRUCTURES[1]

M. Guéron, E. Charretier, M. Kochoyan and J.L. Leroy

Groupe de Biophysique, Ecole Polytechnique, 91128 Palaiseau, France

ABSTRACT Imino proton exchange in duplexes is catalysed by proton acceptors. In the limit of infinite catalyst concentration, the exchange time is equal to the lifetime of the closed base pair. Using spectral broadening, longitudinal relaxation and magnetization transfer in H_2O solutions, exchange times between 1 ms and 1 s can be measured. Those longer than 2 minutes are accessible to direct measurement in D_2O.

Using such NMR methods, we have shown that base pairs open one at a time. In B-DNA, the lifetime is in the range of milliseconds at room temperature, and the influence of neighbors is not large.

In oligodeoxyduplexes containing tracts of A.T base pairs, the lifetimes are anomalously long if the tract contains four consecutive A.T base pairs, possibly including a 5'-AT step but not a 5'-TA step. This strongly suggests that, in such tracts, a conformation distinct from standard B-DNA is formed cooperatively. The lifetime anomaly occurs in all A.T tracts known to generate curved DNA. This is the first local and physical property shown to correlate with DNA curvature.

In the Z form, the lifetime is 100 times longer than in B-DNA.

Exchange kinetics also reveal intrinsic exchange catalysis, and Hoogsteen pairing of G.C pairs.

1. INTRODUCTION

Base-pair opening can be probed by the reactivity of the imino groups of guanosine, thymidine or uridine. Such studies may provide extensive information on base-pair lifetimes, base-pair dissociation constants, the mode of pairing, the accessibility of the open base pair, etc.

One can use chemical probes such as mercury (1) and formaldehyde (2a, 2b) or isotopic probes such as tritium (3, 4, 5) and deuterium (6, 7). Most of the recent work uses proton magnetic resonance to study proton-proton and proton-deuterium exchange (Fig.1), because NMR provides precise identification of the exchanging group (8).

---

[1] This work was supported in part by the Ministère de la Recherche, Action concertée "Structure et Fonction des Macromolécules".

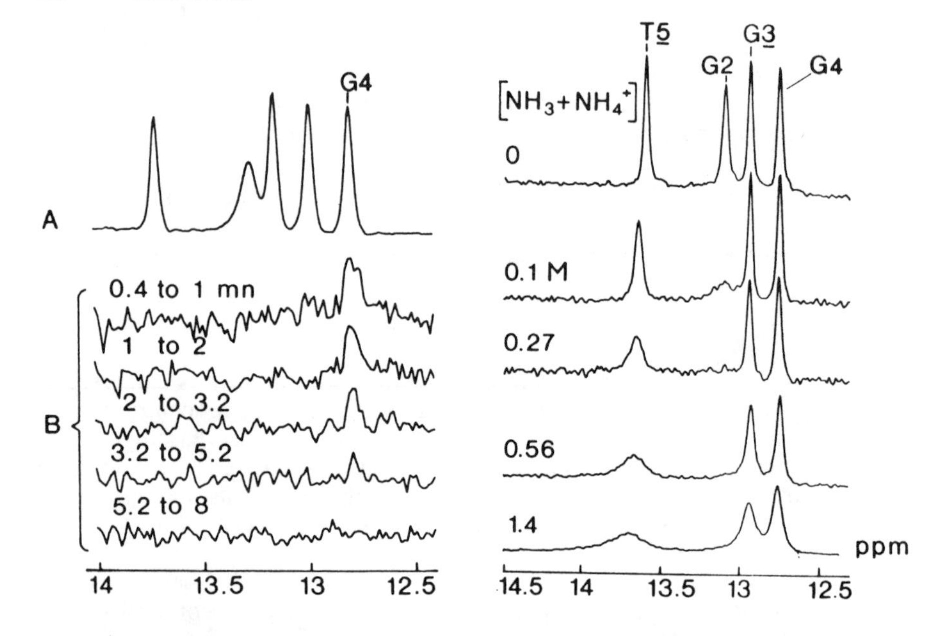

FIGURE 1. Imino proton exchange experiments on the 5'-d(CGCGATCGCG) duplex (T5 means T paired to A5).
Left panel. Real time solvent exchange at pH 7.2, 0°C. (A) reference spectrum in 90% H_2O. (B) difference spectra between the spectra obtained at various times after dilution of the protonated sample in D_2O and the final spectrum at 30 min. The exchange time of the G4 imino proton is 3 minutes.
Right panel. Exchange broadening of the imino protons for increasing concentrations of ammonia. T= 15°C.

Recent NMR studies of synthetic RNA (9), transfer RNA (10), 5s RNA (11) and DNA (8,12) have shown that exchange of imino protons is catalyzed by proton acceptors such as tris, imidazole or ammonia (Figs.1,2). The base-pair lifetimes, derived by extrapolation to infinite catalyst concentration of the imino proton exchange time (Fig.2B), are found to be in the range of milliseconds for RNA and B-DNA duplexes.

Base-pair lifetimes are fairly easy to measure, and they provide interesting probes of nucleic acid structures. When the DNA sequence is random, chemically speaking, so that the structure is probably close to standard B-DNA, the lifetimes are insensitive to sequence, and in particular to the nature of the neighboring base pairs. But stretches of A display anomalously long base-pair lifetimes, which may be related to the so-called B' or P structure (13, 14). This is also the case for GC sequences in the Z form. The B structure itself has its secrets, such as the transient occurence of Hoogsteen G.C pairs, demonstrated and quantitatively characterized by exchange kinetics. Complex molecules, such as tRNA, have also been investigated fruitfully by imino proton exchange (10).

The action of catalysts provides a probe of the open state. The dissociation constant of a base pair may be evaluated by comparison with the exchange rate of the mononucleoside. Imino proton exchange is observed even in the absence of added catalyst, due to exchange catalysis in the open state by the acceptor group of the opposite base, a process taking place through a bridging water molecule (15). Whether opening is to the minor or major grove remains unknown.

In G.C base pairs, the proton exchange rate is enhanced at pH less than 6.5. This remarkable phenomenon is ascribed to the transient formation of Hoogsteen pairs, and these are demonstrated directly at pH 3.

Future applications of proton exchange may include renewed studies of tRNA and studies of other RNAs, studies of triple helices and investigations of complexes of nucleic acids with drugs, proteins or other molecules. Some of these are presently under investigation. Here, we shall consider mainly the applications of imino proton exchange to structural features of DNA duplexes.

2. THEORY

As far as we know, the exchange of imino protons inside a double helix requires the disruption of the base pair, which occurs at a rate $1/\tau_0$, where τ_0 is the base-pair lifetime. Let us take the example of an A.T base pair :

$$\mathrm{TH\text{--}A} \underset{k_{cl}}{\overset{1/\tau_0}{\rightleftharpoons}} \mathrm{TH + A}$$

where k_{cl} is the closing rate. Exchange occurs in the open state via formation of a hydrogen-bonded complex with a catalyst C (a proton acceptor):

$$\mathrm{TH + C \longrightarrow TH\text{--}C \longrightarrow T^- + CH^+}$$

In the open state, the transfer rate k_{tr} of the imino proton is proportional to the catalyst concentration [C] and is a function of the difference between the pK of the nucleoside, pK_N [respectively 9.4 and 9.9 for guanosine and for thymidine at 15°C, (16)], and that of the catalyst, pK_C (17):

$$k_{tr} = k_{coll}\,[C]/(1+10^{(pK_N - pK_C)})$$

where k_{coll} is the collision rate constant. The concentration of the base catalyst [C] is related to the total concentration of buffer [B] by:

$$[C] = [B]/(1+10^{(pK_C - pH)})$$

If the accessibility of the open state were perfect, the transfer rate from the open state, k_{tr}, and from an isolated nucleoside, would be the same.

For a stable base pair, i.e. $\tau_0 k_{cl} >> 1$, the general expression for the exchange time is :

$$\tau_{ex} = 1/k_{ex} = \tau_0 (1+k_{cl}/k_{tr})$$

Since k_{tr} is proportional to the catalyst concentration, a plot of the imino proton exchange time versus the inverse of the catalyst concentration is a straight line which extrapolates to the lifetime of the base pair for infinite catalyst concentration (Fig. 2).

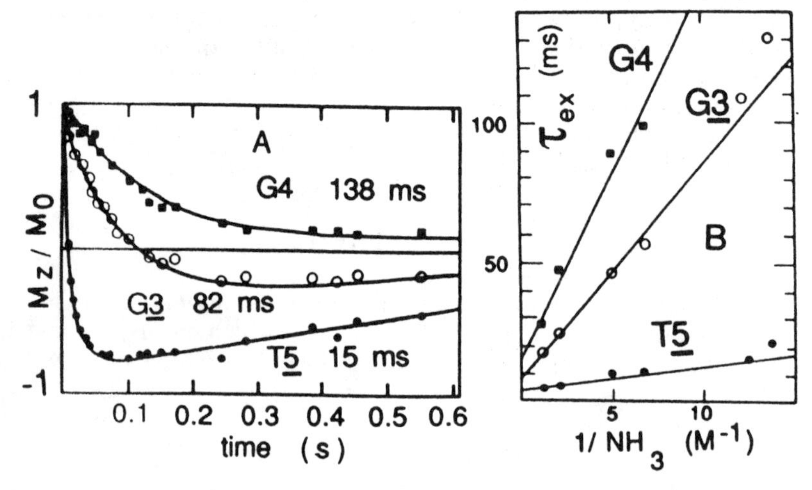

FIGURE 2. *Left panel.* Magnetization transfer experiment (25°C, pH 8.8, 65mM NH_3) on the 5'-d(CGCGATCGCG) decamer. The magnetization of the imino protons is plotted vs. the delay between selective inversion of water and observation. The imino proton exchange time is indicated. It is obtained by curve fitting, in which it is the only adjustable parameter (8).
Right panel. Exchange time at 15°C vs. the inverse of ammonia concentration. Extrapolation to infinite ammonia concentration gives the base-pair lifetime.

Addition of catalyst gives no observable effect in three cases: if the catalyst efficiency is low; if another catalyst, such as an acceptor group of the nucleic acid itself, dominates; or if proton exchange is opening limited.

There is no easy way to distinguish between the three cases. To determine the base-pair lifetime, one must therefore find conditions in which proton exchange is catalyst-dependent.

3. BASE-PAIR LIFETIMES

In this section, we discuss base-pair lifetimes in three different structural situations. The reference is provided by sequences which presumably have a standard B structure. We also consider sequences containing adenine tracts, and lastly turn to Z-DNA. The lifetimes are collected in Table 1.

3.1. B-DNA

The kinetic and exchange properties of the 10-mer self-complementary duplex I, 5'-d(CGCGATCGCG), are representative of those observed in numerous other sequences (8).

The lifetimes are influenced by the nature of pairs: they are shorter for A.T (1 to 7ms at 15°C) than for G.C pairs (7 to 40 ms). But the lifetime of an A.T is not increased by placing it next to a G.C (compare duplexes I and II). Nor is the lifetime of a G.C pair shortened by A.T neighbors.

The lifetimes of neighboring base pairs are different (duplex III). Coupled with the small dissociation constant, 10^{-5} to 10^{-6} (15), this shows that *base pairs open one at a time.*

An interesting feature of the "mirror" sequences, duplexes IV and V, is the large difference in lifetimes of the central G.C pairs (18). These pairs are at the junction between six pyrimidines and six purines. Molecular structures based on 2D NMR have been proposed (19). Both duplexes are B type, but they are distorted in the central region.

In duplex VI, the A.T base-pair lifetime is particularly short. The GTG triplet is commonly found in protein recognition sites (20).

Temperature dependence of base-pair lifetimes in B-DNA. The activation enthalpies range from 40 to 65 kJ/M with no clearcut difference between A.T and G.C pairs. This is the same range as that of the enthalpies for adding a base pair to a double-stranded RNA (21).

But the activation enthalpy of the anomalously short-lived G.C pair in duplex VI is only 30 kJ/mol (8).

3.2. Adenine tracts and their relation to DNA curvature.

The curvature of DNA induced by tracts of A.T base pairs has been studied by the anomalous electrophoretic mobility of duplexes built on repeating decamer sequences (22).It is not yet known if a specific structure of A.T base pairs or of A.T tracts is involved (23, 24).

Consider for example the two decamers 5'-d(CCTTTAAAGG) and 5'-d(GGAAATTTCC) (VII and VIII), which give rise by repetition to straight and curved polymers respectively (25). The difference could result from the geometrical combination of dinucleotide steps among which CT, TA and AG in the first polymer differ from GA, AT and TC in the second,

TABLE 1. **Base-pair lifetimes at 15°C (milliseconds)**

Sequence / lifetimes	Duplex
d-C G C G A : T C G C G	
* * 9 16 4	I
d-C G C G A [T A]$_5$ T C G C G	
(a) (a) (b) (b) (b) 6(a)	II
d-G G [A G]$_6$: [C T]$_6$ C C	
* * 2(a) 19(a)	III
d-C C T T T C : G A A A G G	
* * 4 7 4 **40**	IV
d-G G A A A **G** : C T T T C C	
* * 1 7 7 **7**	V
d-G G [**T** G]$_6$: [C A]$_6$ C C	
* * **0.8** (a) 19(a)	VI
d-C C **T T T** : A A A G G	
* * **≤1 5 1**	VII
d-G G **A A A** : T T T C C	
* * **≤1 86 86**	VIII
d-C G C **A A A A A A** G C G	
* * 12 ≤3 **54 122 91 84 28** 11 * *	IX
d-C G C **I I I I I I** G C G	
* * 6 (c) **2** (a) 6 (c) * *	X
.d-C G C **I** C **I** C **I** C G C G	
* * 3 (c) **0.8** (a) 3 (c) * *	XI
d-C G [C G]$_{10}$ C G ; B form	
* * 7 (a) * *	XII
d-C G [**C G**]$_{10}$ C G ; Z form , NaCl = 4M	
* * **1000** (a) * *	XIII

*Less than 1 millisecond.(a) Unresolved peaks. (b) Same lifetime as the corresponding base pair in duplex I . (c) These two peaks are not resolved.

whereas all the AA steps (and TT steps) are the same in both sequences. Alternatively, the structures of the A_3 (and/or T_3) stretches could differ in the two sequences.

We have found large anomalies in the base-pair kinetics of adenine tracts, in striking correlation with DNA curvature. In the DNA-curving sequence VIII, the lifetimes of the second and third A.T base pairs are extraordinarily long, more than ten times longer than the standard values discussed above. By contrast, the lifetimes are normal in sequence VII, which does not induce curvature.

The comparison of A_3T_3 and T_3A_3 shows that base-pair lifetimes and, by extension, DNA conformation, are dependent on a neighborhood extending beyond first neighbors. *This indicates that DNA structure cannot be described solely by a universal set of dinucleotide geometries.*

The adenine tract of duplex IX also displays anomalous lifetimes and induces DNA curvature. In the crystal structure (13), its A.T base pairs have a large propeller twist, neighboring base pairs are connected by bifurcated hydrogen bonds and the minor groove is narrow. In sequences carrying an A_n tract, we find that, for n=2 and 3, the A.T base-pair lifetimes are normal, i.e. in the range of 1 to 7 ms at 15°C. But for n = 4, the lifetimes are anomalously long, and they increase even more for n = 5, 6 and more.

Figure 3 shows the lifetimes in the A_5 tract. They remain long when I.C is substituted for the third A.T pair. However I.C lifetimes in duplex X (Table 1 and Fig.4) are short. The I.C pair is compatible with a narrow minor groove, but cannot accept or donate the bifurcated hydrogen bond.

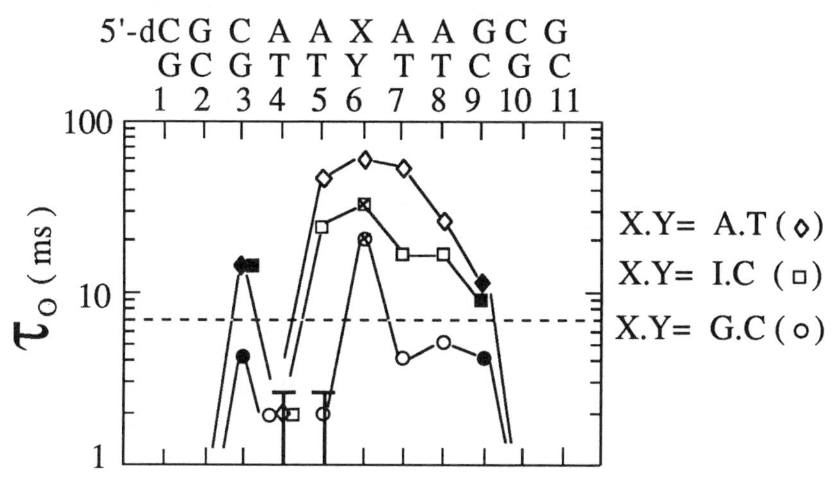

FIGURE 3. Base-pair lifetimes of 5'-d(CGCA$_2$XA$_2$GCG), where X stands for A, I or G, paired with T, C and C respectively. The dotted line signals the longest A.T base-pair lifetime observed in any B-type duplex except for adenine tracts.

The lifetimes are short if the substituent is a G.C pair, which cannot fit in a narrow minor groove.

We have proposed a rule for the occurence of long lifetimes, as follows (26).

Anomalously long base-pair lifetimes are found in DNA tracts of contiguous A.T base pairs. The tract must have a length of at least four (four is marginal), which may include a 5'-AT-3' step, but not a TA step. The A.T base pair at the 5'-A end of the tract is not anomalously long. No G.C pair is part of the tract.

This simple rule is followed without exception by the 17 sequences whose lifetimes have been measured.

Structural implications. The kinetic anomaly is strong evidence for a structure which deviates from B-DNA, forms cooperatively, and requires at least four A.T base pairs. It could be a structure of the type observed for A tracts in crystals (13, 14).

Its biochemical significance is suggested by two observations. First, the disruptive effect of a TA step on lifetimes may be put in relation with the observation that such steps respond differently from other sequences of A.T pairs to attack by nucleases (27). Second, the minimum tract length required for the kinetic anomaly coincides with the minimum tract length (four or five) which displays preferential settings on the nucleosome supercoil (13, 28).

Connection with DNA curvature. Figure 4 correlates the electrophoretic mobility anomaly R of the repeated 10-mer with the longest A.T base-pair lifetime of the tract, in all thirteen cases where the data are available. The correspondance is striking. First, all the tracts with short (≤ 7 ms) lifetimes have normal mobility ($k \ln R \leq 0.18$). The corresponding points fall at the lower left of the plot.

Second, most of the tracts with long lifetimes have a large mobility anomaly ($k \ln R \geq 0.39$), indicative of curvature. The corresponding points fall at the upper right of the plot.

Third, the lower right-hand area is empty: all polydecamers known to be curved are based on sequences displaying long lifetimes.

These observations suggest that the structure causing anomalous base-pair lifetimes is involved in DNA-curving.

The correlation fails only for the three sequences in the upper left-hand corner of Fig.4, corresponding to long lifetimes but no curvature.

3.3. Proton exchange and base pair lifetime in Z-DNA

Base-pair kinetics of Z-DNA have been investigated in the $d(CG)_6$, $d(CG)_{12}$ and $d(C\,^{Br}GCGC\,^{Br}G)$ oligodeoxynucleotides.

The base-pair lifetimes of the CG 12-mer and CG 24-mer (Fig. 5) are similar despite the different lengths, supporting the assumption that exchange occurs via local opening of base pairs and not via a cooperative conversion of the duplex to a non-Z structure (30). The estimated lifetime at

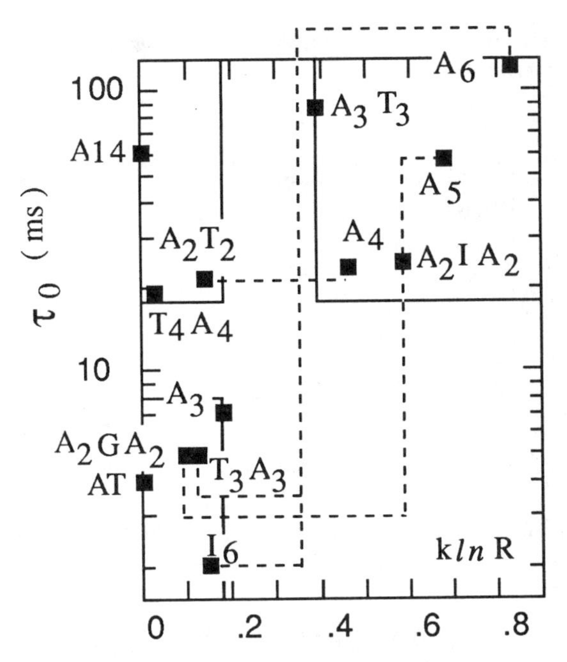

FIGURE 4. Correlation plot of the room temperaure electrophoretic mobility anomaly of different decamer repeats, labelled by their adenine tract, with the longest A.T base-pair lifetime of the duplex at 15°C (26). A point for the tract $d(I_6).d(C_6)$ is also included. The electrophoretic mobility anomaly is characterized by k ln R, where R is the inverse of the ratio of the length of the decamer repeat (150 base pairs) to the length of a random sequence having the same mobility, and k is a correction factor for gel composition (29).

15°C is 800 ms, two orders of magnitude larger than the lifetime of the same base pair in the same duplexes in the B form (Table 1, XII and XIII).

At 5°C, the imino proton exchange times are about 30 min in the absence of added catalyst (Fig. 5). When catalyst was added, the exchange time became less than 100 s. This value provides an upper limit for the base-pair lifetime.

4. MECHANISM of EXCHANGE. The OPEN STATE.

4.1. Exchange in the absence of added catalyst.

The exchange time is surprisingly short in the Absence of Added Catalyst (the AAC exchange process). AAC exchange is too fast to be due to water, it is pH-independent when the pH is not too high (or too low, in the case of guanosine) and it is unaffected by catalysts, below a minimum concentration. The threshold explains why the effect of added catalysts had long been missed.

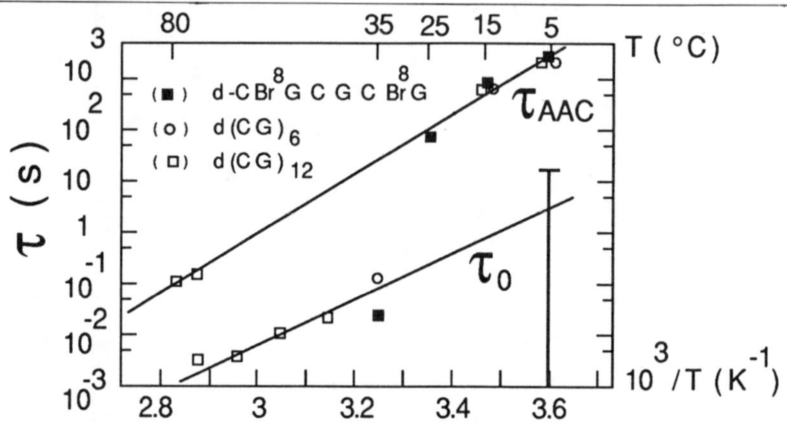

FIGURE 5. Imino proton exchange time, τ_{AAC}, in the absence of added catalyst, and base pair lifetime, τ_0, for a G.C pair of Z-DNA versus the inverse of the temperature.

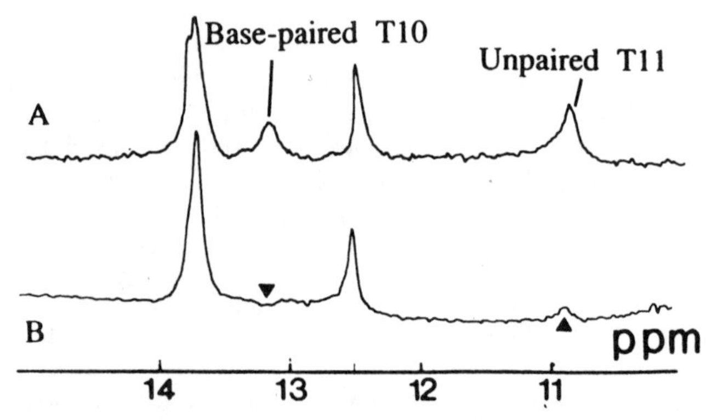

FIGURE 6. Proton exchange in the 5'-d(AATTGCAATTT) undecanucleotide (pH 4.8; T = 15°C).

(A) Reference spectrum. (B) Spectrum obtained 12 ms after selective inversion of water. The magnetization of the imino proton of base-paired T10 is already negative whereas that of unpaired T11 is still positive. The exchange time of the paired imino proton (9 ms) is *shorter* than that of the unpaired one (30 ms)!

The AAC exchange is ascribed to an *intrinsic catalyst* (15). There is extensive indirect evidence for this process, and direct evidence as well (31).

An example is provided by exchange in the duplex formed by the undecamer 5'-d(AATTGCAATTT), in which the last T is single whereas the penultimate T is base-paired (Fig. 6). Nevertheless, the imino proton of the penultimate T is base-paired (Fig. 6). Nevertheless, the imino proton of the latter exchanges *faster*, hence by a mechanism which is inaccessible to the unpaired T (Fig. 8). We have shown (15) that catalysis in the open state by N1 of adenine of the same base pair can provide the mechanism.

Due to the weak catalytic efficiency of this group (the pK of adenosine N_1 is 3.5), the two bases must remain close to each other. A model of the open state where one water molecule (or maybe more) connects the thymidine imino proton to the adenosine N_1 explains both the net proton transfer to water by the internal catalyst, and the large accessibility of added catalysts to the imino proton. The same process should also apply to the open state of G.C pairs.

4.2. Acid catalysis of proton exchange in G.C base pairs. Hoogsteen base pairs in DNA duplexes.

For G.C base pairs, imino proton exchange is acid-catalysed at pH<6.5. The process is extraordinarily fast, as fast as the acid-catalysed exchange computed for isolated guanosine. An explanation was searched for, in the formation of a CH^+.G Hoogsteen base pair, a structure in which the guanosine proton is exposed, and whose stability is pH dependent, since it requires protonation of cytidine.

In $d(CG)_{12}$ at pH 3.2, a resonance appears at 15.5 ppm. It is assigned by its NOE to guanosine H8, to the imino proton of CH^+ in a Hoogsteen base pair. Its intensity increases at lower pH, while that of the G

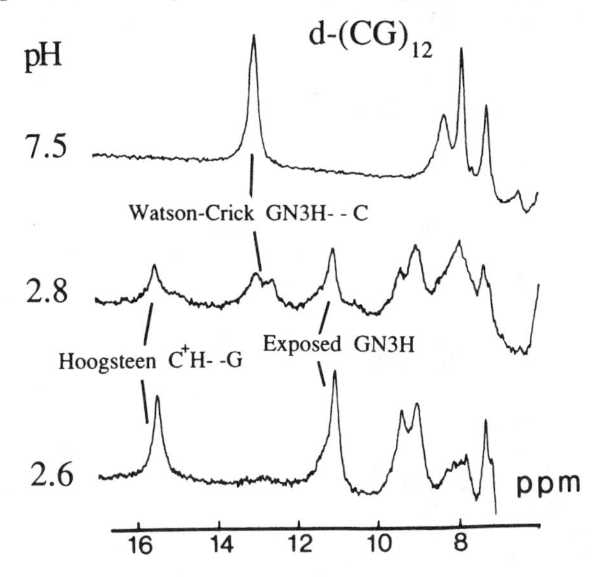

FIGURE 7. Proton NMR spectra of the $d(CG)_{12}$ duplex versus pH. T=0°C, 100 mM $NaClO_4$, 10% D_2O.

imino proton of the Watson-Crick base pair decreases. At pH 2.6, it corresponds to 70% of the base pairs (Fig. 7).

The Hoogsteen to Watson-Crick equilibrium constant should vary as 10^{-pH}, leading to a value ca. 10^{-4} at pH 7. These observations provide strong evidence for the existence of CH^+.G Hoogsteen pairs in DNA, in agreement with an earlier study (32), and even at neutral pH as a minor species. One should now consider whether a small fraction of the A.T base pairs of DNA duplexes might also exist in the Hoogsteen form.

REFERENCES

1. Williams MN, Crothers DM (1975). Binding kinetics of mercury(II) to polyribonucleotides. Biochemistry 14:1944.

2a. McGhee JD, von Hippel PH (1977). Formaldehyde as a probe of DNA structure. 3. Equilibrium denaturation of DNA and synthetic polynucleotides. Biochemistry 16:3276.

2b. Franck-Kamenetskii MD (1985). Fluctuational motility of DNA. In Clementi E, Corongin G, Sarma MH, Sarma H (eds): "Structure and motion: membranes, nucleic acids and proteins." New York: Adenine Press, Guiderland, p 417.

3. Printz MP, von Hippel PH (1965). Hydrogen exchange studies of DNA structure. Proc Nat Acad Sci U.S.A. 53:363.

4. Teitelbaum H, Englander SW (1975). Open states in native polynucleotides. I. Hydrogen-exchange study of adenine-containing double helices. J Mol Biol 92:55.

5. Teitelbaum H, Englander SW (1975). Open states in native polynucleotides. II. Hydrogen-exchange study of cytosine-containing double helices. J Mol Biol 92:79.

6. Nakanishi N, Mitane Y, Tsuboi M (1984). A hydrogen exchange study of the open segment in a DNA double helix. Biochim Biophys Acta 798:46.

7. Hartmann B, Leng M, Ramstein J (1986). Poly(dA-dT).poly(dA-dT) two-pathway proton exchange mechanism. Effect of general and specific base catalysis on deuteration rates. Biochemistry 25:3073.

8. Leroy JL, Kochoyan M, Huynh-Dinh T, Guéron M (1988). Characterization of base-pair lifetimes in deoxy-duplexes using catalysed exchange of the imino proton. J Mol Biol 200:223.

9. Leroy JL, Broseta D, Guéron M (1985). Proton exchange and base-pair kinetics of poly(rA).poly(rU) and poly(rI).poly(rC). J Mol Biol 184:165.

10. Leroy JL, Bolo N, Figueroa N, Plateau P, Guéron M (1985). Internal motions of transfer RNA: a study of exchanging protons by magnetic resonance. J Biomol Struct Dynam 2:915.

11. Leontis NB, Moore PB (1986). Imino proton exchange in the 5S RNA of *Escherichia coli* and its complex with protein L25 at 490 Mhz. Biochemistry 25:5736.

12. Bendel P (1987). A proton NMR spin-lattice relaxation study of the imino proton exchange kinetics in calf-thymus DNA. Biopolymers 26:573
13. Nelson HCM, Finch JT, Luisi BF, Klug A (1987). The structure of an oligo(dA).oligo(dT) tract and its biological implications. Nature 330:221.
14. Coll M, Frederick CA, Wang AH-J, Rich A (1987). A bifurcated hydrogen-bonded conformation in the d(A.T) base pairs of the DNA dodecamer d(CGCAAATTTGCG) and its complex with distamycin. Proc Natl Acad Sci USA 84:8385.
15. Guéron M, Kochoyan M, Leroy JL (1987). A single mode of DNA base-pair opening drives imino proton exchange. Nature 382:89.
16. Izatt RM, Christensen JJ, Rytting JH (1971). Sites and thermodynamic quantities associated with proton and metal ion interaction with ribonucleic acid, deoxyribonucleic acid, and their constituent bases, nucleosides and nucleotides. Chem Rev 71:439.
17. Eigen M (1964). Proton transfer, acid-base catalysis, and enzymatic hydrolysis. Angew Chem Int Ed Engl 3:1.
18. Kochoyan M, Leroy JL, Guéron M (1987). Proton exchange and base-pair lifetimes in a deoxy-duplex containing a purine-pyrimidine step and in the duplex of inverse sequence. J Mol Biol 196:599.
19. Patel DJ, Shapiro L, Hare D, (1987).DNA and RNA: nmr studies of conformations and dynamics in solution. Quart. Rev. Biop. 20:35.
20. Donlan M, Cheung S, Lu P (1986). Nuclear magnetic resonance studies of an SV40 enhancer core DNA sequence. Biophys J 49:26.
21. Borer PN, Dengler B, Tinoco I Jr (1974). Stability of ribonucleic acid double-stranded helices. J Mol Biol 86:843.
22. Koo HS, Wu HM, Crothers DM (1986). DNA bending at adenine-thymine tracts. Nature 320:501.
23. Wu HM, Crothers D. (1984). The locus of sequence-directed and protein-induced DNA bending. Nature 308:509.
24. Diekman S, Von Kintzing E (1988). DNA bending and curvature. In Olson MH, Sarma MH ,Sarma RH, Sundaralingam M,(eds): "Structure and Expression, Vol.3" New York: Adenine Press, Guilderland, p 57.
25. Hagerman PJ (1988). Sequence-dependent curvature of DNA. In Wells RD, Harvey SC (eds):"Unusual DNA Structures" New York: Springer-Verlag, p 225.
26. Leroy JL, Charretier E, Kochoyan M, Guéron M (1988). Evidence from base-pair kinetics for two types of adenine tract structures in solution: their relation to DNA curvature. Biochemistry 27:8894.
27. Drew HR, Travers AA (1984). DNA structural variations in the E.coli *tyr*T promoter. Cell 37:491.
28. Satchwell SC, Drew HR, Travers AA (1986). Sequence periodicities in chicken nucleosome core DNA. J Mol Biol 191:659.
29. Calladine CR, Drew HR, McCall MJ (1988). The intrinsic curvature of DNA in solution. J Mol Biol 201:907.
30. Mirau PA, Kearns DR (1985).Unusual proton exchange properties of Z-form poly[d(G-C)]. Proc Natl Acad Sci USA 82:1594.
31. Kochoyan M, Lancelot G, Leroy JL (1988). Study of structure, base-pair opening kinetics and proton exchange mechanism of the d-(AATTGCAATT)

self-complementary oligodeoxynucleotide in solution. Nucl Acid Res 16:7685
32. Courtois Y, Fromageot P, Guschlbauer W (1968). Protonated polynucleotide structures. 3.An optical rotatory dispersion study of the protonation of DNA. Biochemistry 6:493.

Frontiers of NMR in Molecular Biology, pages 239-248

STRUCTURE AND DYNAMICS OF DISTAMYCIN A WITH d(CGCAAATTGGC)·d(GCCAATTTGCG)[1]

Jeffrey G. Pelton and David E. Wemmer

Department of Chemistry, University of California, and Chemical Biodynamics Division, Lawrence Berkeley Laboratory, 1 Cyclotron Road, Berkeley, California 94720

ABSTRACT The binding of distamycin A to d(CGCAAATTGGC)·d(GCCAATTTGCG) was studied by two-dimensional NMR at both low and high drug/DNA ratios. At low drug ratios the drug binds to a family of minor-groove A-T binding sites. At high drug ratios two drugs bind simultaneously in the minor groove of each undecamer. They are in complete van der Waals contact with each other, and are oriented with the positive charge of each pointing toward an opposite end of the helix. The drug exchanges among all of the binding sites.

INTRODUCTION

The efficient design of new drugs is based on a knowledge of both the mechanism of action and structures of existing drugs and their complexes with relevant biological macromolecules. Distamycin A (Figure 1) is an antibiotic that binds to the minor groove of A-T rich B-form DNA sequences (1). It inhibits RNA synthesis by sterically blocking RNA polymerase through binding with A-T rich initiation sites. Although the mechanism of

---

[1]This work was supported by NIH through the U.C.B. BRSG program and equipment grants from the DOE University Research instrumentation program (DE F605 86ER75281) and NSF (DMB 8609035).

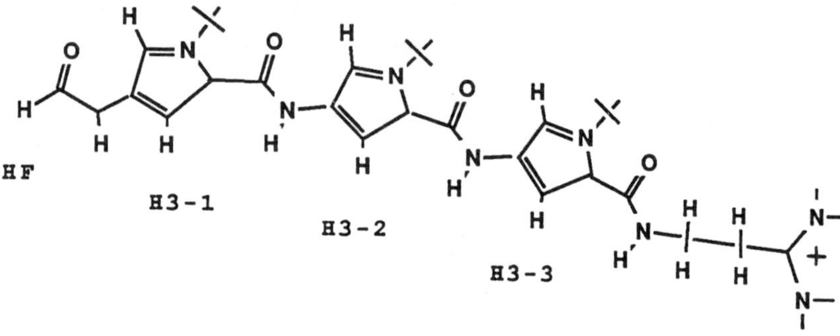

Figure 1. Schematic of Distamycin A. The drug consists of a formyl group, followed by three N-methylpyrrole carboxamide groups, and a positively charged propylamidinium group.

distamycin A's action is understood, the specificity of the drug for different A-T sites remains unclear.

In a study using affinity cleaving and footprinting methods, distamycin A formed a complex with the five A-T base sequence 5'-AAATT-3'(2). In a crystallographic study of distamycin A with d(CGCAAATTTGCG)$_2$ (3) only one binding site for the drug was observed, although several sites were available. NMR studies of distamycin A with d(CGCGAATTCGCG)$_2$ (4,5) also revealed one binding site for the drug indicating that a binding site requires only four A-T base pairs. In fact, the binding constant for distamycin A with d(CGCAATTGCG)$_2$ was found to be 2.7×10^8 (6). Similar results have been obtained with netropsin, an analogous minor-groove binding drug (7,8).

We have studied the binding of distamycin A with d(CGCAAATTGGC)·d(GCCAATTTGCG) at the molecular level to gain insight into the specificity of the drug for different A-T sites and to probe the forces responsible for its binding.

Nuclear Overhauser effect spectroscopy (NOESY) (9) two-dimensional NMR experiments contained both nuclear Overhauser effect (NOE) cross peaks that provided information on drug-DNA contacts, and chemical exchange cross peaks that provided information on the dynamics of the system.

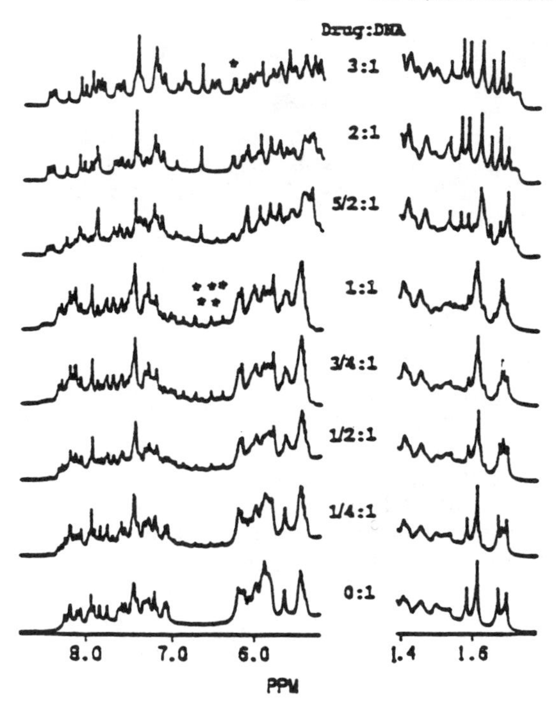

Figure 2. Titration of d(CGCAAATTGGC) with Distamycin A. Stars indicate bound-drug H3 pyrrole resonances.

RESULTS

The result of adding increasing amounts of distamycin A to the undecamer are presented in Figure 2. Up to one equivalent of added drug, two sets of three peaks (relative intensity 1.5/1) appear between 6.3 and 6.7 ppm characteristic of the three drug H3 pyrrole protons (see Figure 1).

Up to two equivalents of added drug, the original six peaks disappear, while two new peaks of equal intensity grow in near 6.3 ppm. These peaks correspond to two forms of the drug H3-2 proton (see below). In addition, five methyl resonances corresponding to the five thymine residues are clearly resolved near 1.6 ppm.

At three equivalents of added drug, free H3 and H5 drug pyrrole resonances appear at 6.46 and

6.83 ppm respectively, indicating that the binding sites have been saturated.

The data at low drug/DNA ratios suggests that the drug binds to two sites on the undecamer, and, taken together, the data at high drug/DNA ratios suggest that two drugs bind in specific binding sites simultaneously.

One-Half Equivalent of Added Drug

An expansion of the aromatic region of a NOESY spectrum at one-half equivalent of added drug is presented in Figure 3. In this spectrum chemical exchange cross peaks correlate each major form drug H3 pyrrole proton, assigned through intramolecular 1D NOEs to the drug amide protons, with the corresponding minor form H3 proton. This indicates that the drug exchanges between binding sites at a rate that is slow on the NMR time scale. Chemical exchange cross peaks also connect each adenine C2H resonance of the free undecamer with two new down-field shifted resonances that correspond to the chemical shifts of these C2H protons upon binding of the drug in one of two sites.

In addition to chemical exchange cross peaks, the NOESY spectrum of Figure 3 contains NOE cross peaks that connect both the major and minor form H3 protons with a bound adenine C2H resonance. These contacts are listed in Table 1.

TABLE 1
Drug-DNA Contacts
at One-Half Equivalent of Added Drug

Drug Proton	DNA Proton
H3-1 M	A4 M, A5 M
H3-2 M	A5 M, A6 M
H3-3 M	A6 M, A16 M
H3-1 m	A6 m and/or A16 m, A15 m
H3-2 m	A6 m and/or A16 m
H3-3 m	A5 m, A6 and/or A16 m

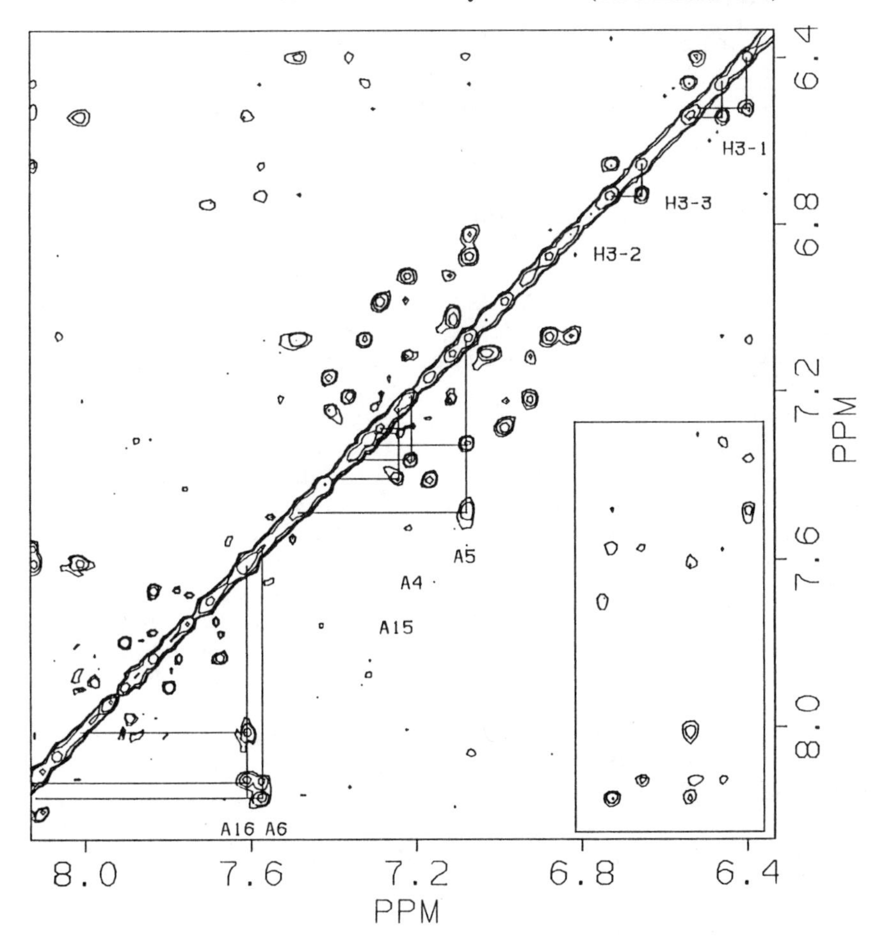

Figure 3. NOESY Spectrum of the Distamycin A - d(CGCAAATTGGC) Complex. Expansion of the aromatic region of a NOESY spectrum of d(CGCAAATTGGC) with one-half equivalent of distamycin A, taken at 35°C. Drug-DNA NOE cross peaks are outlined in the box.

From the intensities of the adenine C2H exchange cross peaks, we estimate an off-rate constant of 2.4/s for the drug in the major binding site, and 3.3/s for the drug in the minor binding site at 35°C, and in 10mM sodium phosphate and 10mM sodium chloride. The drug exchange peaks reflect primarily exchange of the drug between DNA molecules rather than exchange on the same undecamer.

Three Equivalents of Added Drug

An expansion of the DNA aromatic to C1'H region of a NOESY spectrum taken at three equivalents of added drug is presented in Figure 4. In this spectrum chemical exchange cross peaks correlate each free drug H3 proton with two new upfield resonances that correspond to the chemical shifts of these protons in two bound forms. NOE cross peaks also correlate the DNA aromatic and C1'H protons allowing for assignment of these resonances by sequential connectivity methods. In addition to intramolecular DNA-DNA NOEs, intermolecular drug-DNA NOEs are observed between H3 and C2H protons. The drug-drug (NOESY spectrum not shown) and drug-DNA contacts are listed pictorially in the lower portion of Figure 5.

Based on the intensities of the free/bound drug exchange cross peaks, we estimate an off-rate constant of 1/s for the drug exchanging from the two-drug complex to the free form.

At one and one-quarter equivalents of added drug, a NOESY spectrum obtained at 35°C (not shown) revealed exchange, as evidenced by the drug H3 resonances, between the two binding sites observed at low drug/DNA ratios, and the two-drug complex observed at high drug/DNA ratios. Exchange rate constants for these processes ranged from 2-5/s.

DISCUSSION

Previous NMR studies of the distamycin A - d(CGCGAATTCGCG)$_2$ complex revealed that each drug H3 proton gives an NOE, and is in van der Waals contact with only one adenine C2H proton. Two models can account for the present data, in which NOEs are observed between each major form H3 proton and two adenine C2H protons (see Table 1). Either the drug is positioned on the sequence so each H3

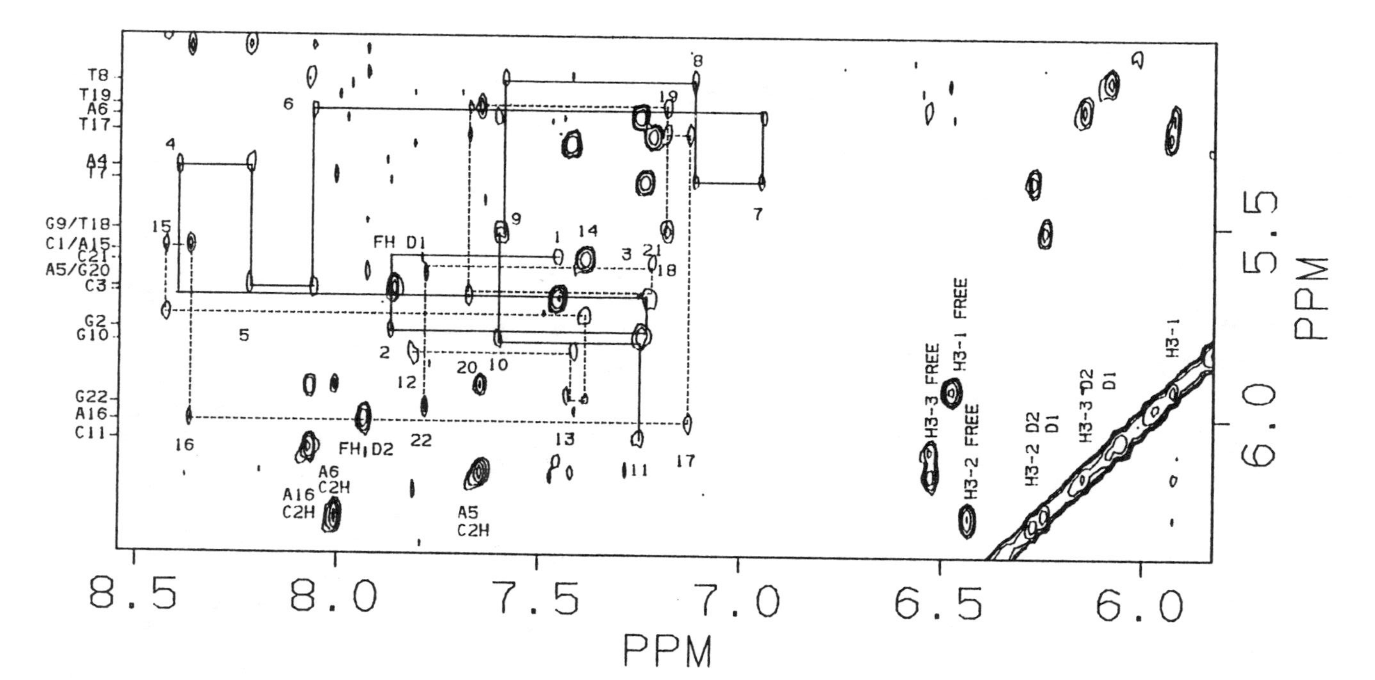

Figure 4. NOESY Spectrum of the Distamycin A - DNA Complex. Expansion of the aromatic to C1'H region of a NOESY spectrum taken at three equivalents of added drug, showing DNA sequential connectivities and drug-DNA contacts. The solid line denotes the sequential connectivities along the A-rich strand, and the dashed line denotes connectivities along the T-rich strand. Aromatic assignments are denoted by numbers, and DNA C1'H assignments are denoted along the vertical axis.

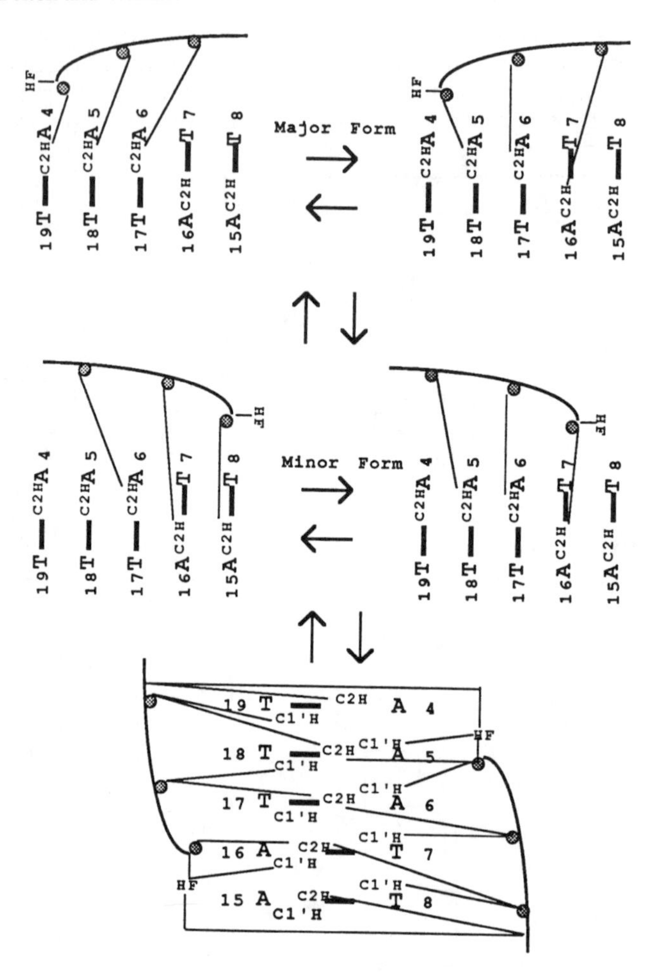

Figure 5. Schematic of Distamycin A - DNA Binding Modes. Lines represent drug - DNA contacts observed in NOESY spectra. The top two sets of two binding modes exist in the presence of substoichiometric amounts of drug. The binding sites in each set are related by a frame shift of the drug by one base pair (fast exchange), while the two sets of sites are related by a flip of the drug with respect to the DNA (slow exchange). The lower structure exists when more than one equivalent of drug is present. Circles represent drug H3 protons.

proton is close to two C2H protons simultaneously, or the drug must slide in fast exchange between two binding sites of the type previously seen (4,5). If a fast exchange model is assumed for both the major and minor binding forms, two sets of two binding sites can be constructed with the drug binding to 5'-AAAT-3', AATT, and ATTT (Figure 5). We favor this sliding model since 5'-AATT-3'(4,5) and 5'-AAAT-3'(10) have been individually characterized as good binding sites previously by NMR. Although the drug covers five base pairs, this model suggests that a distamycn A binding site requires only four A-T base pairs.

The NOE patterns observed in the NOESY spectrum with three equivalents of added drug suggest that two drugs bind simultaneously in the minor groove and are in complete van der Waals contact. This result underscores the importance of intra- and intermolecular stacking interactions in stabilizing the complex, and suggests the possibility of A-T sequence dependent DNA flexibility. In the crystal structure of $d(CGCGAATTCGCG)_2$ (11), the minor groove width of the A-T region, less two phosphate radii, was approximately 4Å. If the thickness of the drug pyrrole rings is taken to be 3.4Å, binding two drugs must expand the minor groove to at least 6.8Å. Whether this binding phenomenon is characteristic of only this sequence, or is a general result common to other A-T rich DNA oligomers is presently under investigation.

ACKNOWLEDGEMENTS

We greatly appreciate the continued collaboration with Professor R. Klevit, and thank D. Koh for help with DNA synthesis.

REFERENCES

1. Hahn FE (1975). Distamycin A and Netropsin. In Corcoran JW, Hahn FE (eds): "Antibiotics III, Mechanism of action of Antimicrobial and Antitumor agents," New York: Springer-Verlag, p79.

2. Schultz PG, Dervan PB (1984). Distamycin and Penta-N-Methylpyrrole Carboxamide Binding Sites on Native DNA, A Comparison of Methidiumpropyl-EDTA-Fe(II) Footprinting and DNA Affinity Cleaving. J Biomol Struct Dyn 1:1133.
3. Coll M, Frederick CA, Wang H-JA, Rich A (1987). A Bifurcated Hydrogen-Bonded Conformation in the d(A·T) Base Pairs of the DNA Dodecamer d(CGCAAATTTGCG) and its Complex with Distamycin. Proc Natl Acad Sci 84:8385.
4. Klevit RE, Wemmer DE, Reid BR (1986). ^{1}H NMR Studies on the Interaction Between Distamycin A and a Symmetrical DNA Dodecamer. Biochemistry 25:3296.
5. Pelton JG, Wemmer DE (1988). Structural Modeling of the Distamycin A-d$(CGCGAATTCGCG)_2$ Complex using 2D NMR and Molecular Mechanics. Biochemistry 27:8088.
6. Breslauer KJ, Remeta DP, Chou WY, Ferrante R, Curry J, Zaunczkowski D, Snyder JG, Marky LA (1987). Enthalpy-Entropy Compensations in Drug-DNA Binding Studies. Proc Natl Acad Sci USA 84:8922.
7. Patel DJ, Shapiro L (1985). Sequence Dependent Recognition of DNA Duplexes. J Biol Chem 261:1230.
8. Kopka ML, Yoon C, Goodsell D, Pjura P, Dickerson RE (1985). Binding of an Antitumor Drug to DNA, Netropsin and CGCGAATT^{Br}CGCG. J Mol Biol 183:553.
9. Macura S, Ernst RR (1980). Elucidation of Cross Relaxation in Liquids by Two-Dimensional NMR Spectroscopy. Mol Phys 41:95.
10. R. Klevit (personal communication)
11. Fratini AV, Kopka ML, Drew HR, Dickerson RE (1982). Reversible Bending and Helix Geometry in a B-DNA Dodecamer. J Biol Chem 257:14686.

Frontiers of NMR in Molecular Biology, pages 249-258

DNA STRUCTURES WITH HOOGSTEEN BASE PAIRS[1]

Juli Feigon[2], Ponni Rajagopal, and Dara Gilbert

Department of Chemistry and Biochemistry and the Molecular Biology Institute, University of California, Los Angeles, CA 90024

ABSTRACT NMR data is presented on two DNA structures which contain Hoogsteen base pairs, (1) a triple-stranded DNA and (2) an echinomycin-DNA complex. We have investigated the structures formed from the homopurine: homopyrimidine sequences d(GAGAGAGA) and d(TCTCTCTC). Under appropriate conditions (pH < 6, excess $d(TC)_4$) the predominant conformation observed is a triple-stranded helix in which a $d(GA)_4$ and a $d(TC)_4$ strand are Watson-Crick base paired and a second $d(TC)_4$ strand is Hoogsteen base paired in the major groove to the $d(GA)_4$ strand. We have also investigated the structure of the complex of echinomycin with d(ACGTACGT). Crystal structures of echinomycin-DNA complexes showed Hoogsteen base pairs formed adjacent to echinomycin binding sites. We have found that Hoogsteen base pairs are formed in the echinomycin-DNA complex in solution at 1°C. However, Hoogsteen base pairs which form in the interior of the DNA duplex are much less stable than those formed at the ends of the duplex, and at physiological temperatures the former are exchanging between the Hoogsteen base paired and an open state.

---

[1]This work was supported by grants from NIH (RO1 GM 37254-02) and ONR (N00014-88-K-0180) to J.F.

[2]Author to whom reprint requests should be addressed.

INTRODUCTION

Alternatives to Watson-Crick base pairing of nucleotides, such as Hoogsteen base pairs, have been recognized since the first crystal structures of nucleotides[1], but until recently they have been thought to be biologically irrelevant for DNA. In this paper, we review NMR evidence for two different DNA structures which contain Hoogsteen base pairs, a triple-stranded DNA and a drug-DNA complex.

TRIPLE-STRAND FORMATION FROM $d(GA)_4$ AND $d(TC)_4$

Runs of homopurine:homopyrimidine sequences are dispersed throughout the eucaryotic genome and appear to have diverse roles in genetic regulation (reviewed in 2). Such sequences show a pH and supercoil density dependent hypersensitivity to DNAse I and single strand specific nucleases[2,3,4]. The most widely accepted model to explain this data is formation of a homopyrimidine:homopurine: homopyrimidine triplex with a single stranded homopurine loop[5]. Evidence for *in vitro* formation of DNA triplexes formed by binding of a homopyrimidine DNA sequence to the complementary double-stranded DNA sequence in a restriction fragment has also recently been presented[6].

We have used one- and two-dimensional ^1H NMR spectroscopy to study the complexes formed between the homopurine d(GAGAGAGA) and the homopyrimidine d(TCTCTCTC)[7,8]. One dimensional ^1H NMR spectra of the base resonances are shown in figure 1. At pH 7.3 and 1:1 $d(GA)_4$:$d(TC)_4$ the predominant complex formed is a B-DNA duplex (figure 1a). Dramatic changes in the spectra are observed when the pH is lowered to 5.5 and excess $d(TC)_4$ strand is added (figure 1b). Under these conditions, the predominant complex is a triple helix composed of one homopurine and two homopyrimidine strands. Resonance assignments for both the duplex and the triplex are indicated on the figure and are described elsewhere[7,8]. Large upfield shifts are observed for the base proton resonances in the

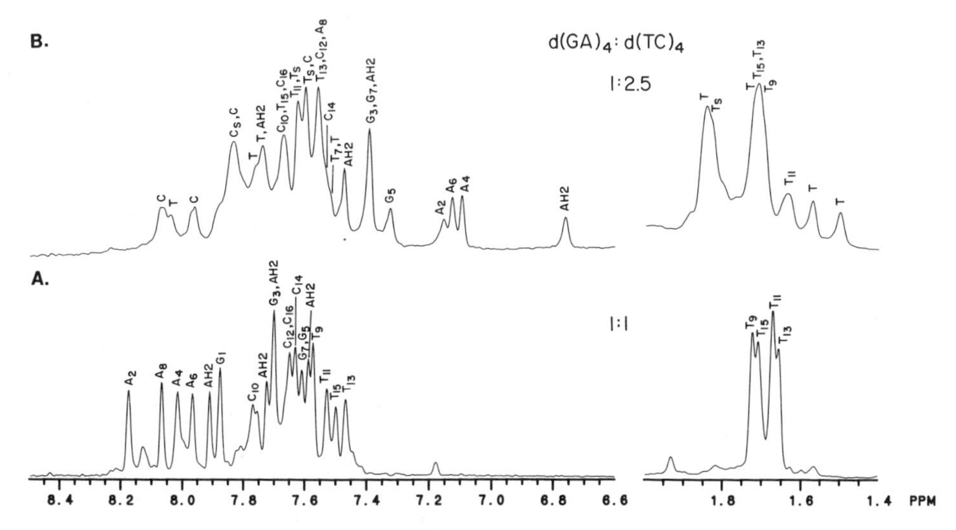

FIGURE 1. 500 MHz ^1H NMR spectra of the aromatic and methyl regions of $d(GA)_4$ + $d(TC)_4$ in D_2O. (A) 1:1 $d(GA)_4$:$d(TC)_4$ in 20 mM phosphate, pH 7.3, 50 mM NaCl, 5 mM $MgCl_2$, 15°C. (B) 1:2.5 $d(GA)_4$:$d(TC)_4$ in 10 mM phosphate, pH 5.5, 100 mM NaCl, 5 mM $MgCl_2$, 35°C.

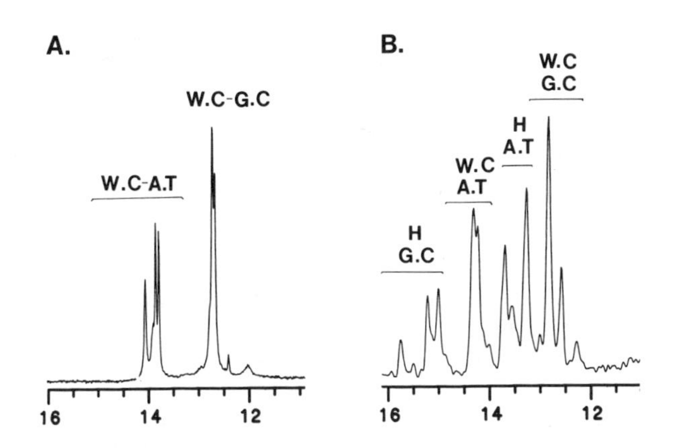

FIGURE 2. 500 MHz ^1H NMR spectra of the imino proton resonances of $d(GA)_4$ + $d(TC)_4$ in H_2O. Samples are the same as in figure 1a and b. (A) 5°C (duplex). (B) 10°C (triplex). Resonances from Watson-Crick (W.C.) and Hoogsteen (H) are identified.

homopurine strand relative to duplex DNA, with three of the adenines showing shifts of greater than 1 ppm.

Imino proton spectra of the same complexes in H_2O are shown in figure 2. At pH 7.1 and 1:1 $d(GA)_4$:$d(TC)_4$ (figure 2a) four GC and four AT imino resonances are observed as expected for a B-DNA duplex. At pH 5.5 and 1:2.5 $d(GA)_4$:$d(TC)_4$ (figure 2b) several additional imino proton resonances are observed and these are indicative of triplex formation. These additional imino resonances have been identified as arising from 4 TA and 3 CG Hoogsteen base pairs which are formed between the second $d(TC)_4$ strand and the $d(GA)_4$ strand in the triplex on the basis of crosspeaks observed in NOESY spectra in water. All of the Hoogsteen base pairs show strong imino-H8 crosspeaks while Watson-Crick base pairs show strong imino-AH2 or G amino crosspeaks (not shown). The base triplets are illustrated below (figure 3) Observation of

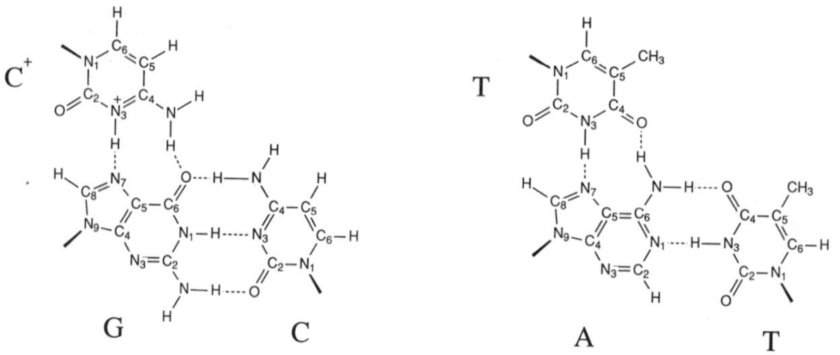

FIGURE 3. C^+-G-C and T-A-T base pairing scheme in the triple strand.

crosspeaks between imino and H8 protons in CG base pairs confirms that the C is protonated at N3 and explains the pH dependence of the triplex formation. Sequential connectivities between the imino protons in the Hoogsteen base paired $d(TC)_4$ strand are shown in figure 3a. We note that the predominant triplex formed has 4 TAT and 3 C^+GC base triplets, but there is a small amount of triplex

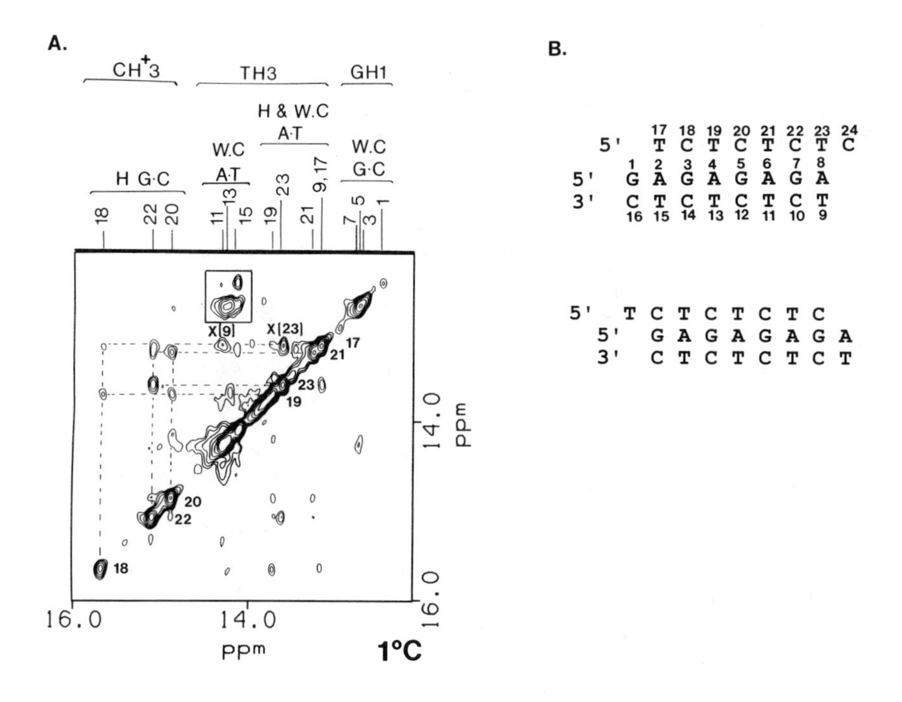

FIGURE 4. (A) Portion of the NOESY spectrum of the triplex in H_2O at 1°C (τ_m = 150 msec) showing the region containing the imino proton resonances and their crosspeaks. Sample is the same as figure 2b. Sequential connectivities between imino protons in the Hoogsteen base pairs are indicated by dashed lines. The boxed region contains crosspeaks between Watson-Crick base paired imino protons. Assignments are indicated at the top of the spectrum. Exchange crosspeaks are marked (x). (B) Triplexes with numbering system. The triplex at top is the predominant structure observed.

with 3 TAT and 4 $C^{+}CG$ triplets (figure 3b) which shows exchange crosspeaks with the major form of the triplex.

To our knowledge this is the first NMR evidence for triplex formation in homopurine:homopyrimidine DNA sequences, and the first direct observation of a protonated C imino in DNA triple helices. One $d(GA)_4$ and one $d(TC)_4$ strand are Watson-Crick base paired, and the second $d(TC)_4$ strand is Hoogsteen base paired in the major groove. The sugars of both of the homopyrimidine strands are in an A-DNA conformation.

ECHINOMYCIN-d(ACGTACGT) COMPLEX

Echinomycin is a cyclic octadepsipeptide antibiotic which bis-intercalates into DNA[9,10]. The preferred binding site contains a central CpG step[11,12]. Crystal structures of echinomycin complexed to DNA revealed that the base pairs adjacent to the bis-intercalation site were in the Hoogsteen conformation[13-15]. We have studied the solution conformation of the $d(ACGTACGT)_2$-echinomycin complex using ^1H and ^{31}P NMR[16,17]. Two echinomycin molecules are found to bind per duplex as illustrated schematically below. The binding is cooperative and symmetric.

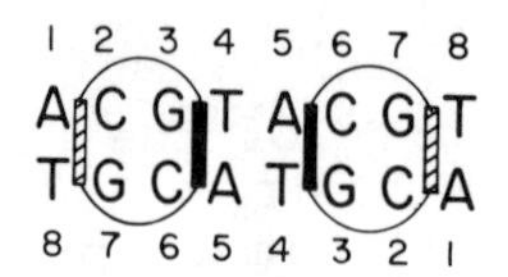

Stacked plots of NOESY spectra of the DNA-echinomycin complex are shown in figure 5 as a function of temperature. At 1°C, strong crosspeaks are observed between both the terminal and the internal AH8 resonances and their H1' sugar resonances. This indicates that these bases are in the syn conformation as expected if they are Hoogsteen base paired in this structure[16]. In contrast, the other bases are all anti as is usual for B-DNA[18]. Confirmation of Hoogsteen base pairing for the internal AT base pairs is obtained from the NOESY spectra in H_2O at 1°C (not shown), which show AT imino-AH8-AH1' crosspeak connectivities (figure 6).

Although both the terminal and internal AT base pairs in the echinomycin-d(ACGTACGT) complex are in the Hoogsteen conformation at low temperatures, the

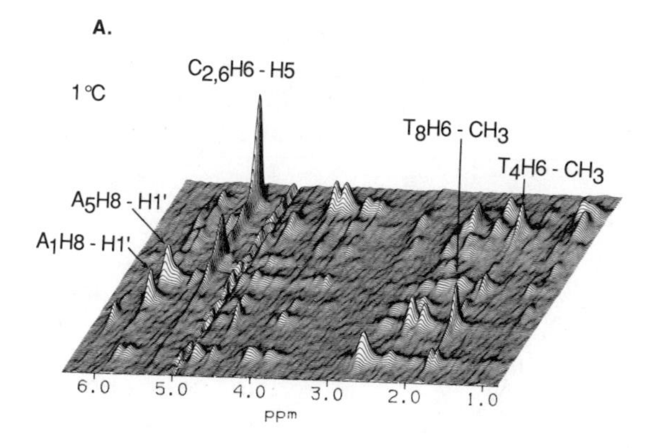

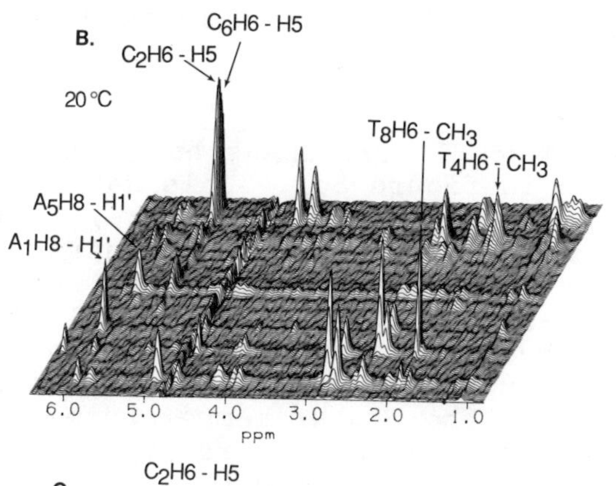

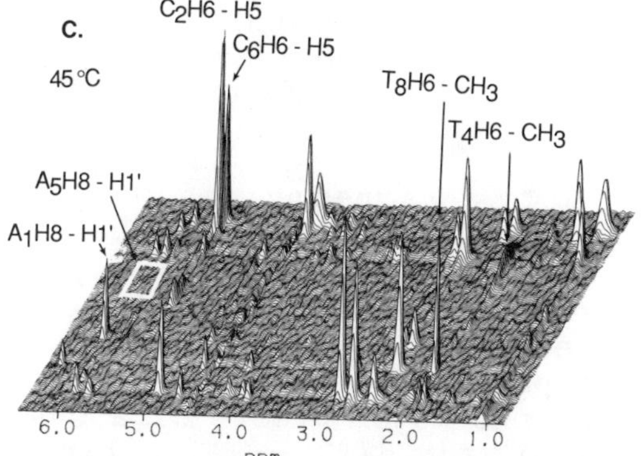

FIGURE 5. Stacked plots of NOESY spectra of the 2 echinomycin:[d(ACGTACGT)]$_2$ complex in D_2O as a function of temperature. (A) 1°C, τ_m=50 msec, (B) 20°C, τ_m=100 msec, (C) 45°C, τ_m=180 msec. The region of crosspeaks between the aromatic and the DNA sugar, H5, CH_3, and drug peptide resonances is shown.

conformation begins to change as the temperature is raised. At 20°C (figure 5b), the A_4H8-H1' crosspeak has decreased in intensity, while the terminal A_1H8-H1' crosspeak remains as intense as it is at 1°C.

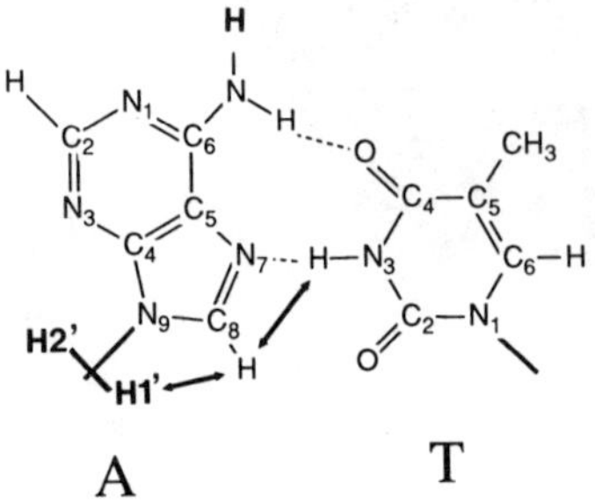

FIGURE 6. AT Hoogsteen base pair with the adenine in the *syn* conformation. Arrows indicate the short interproton distances expected to give rise to large NOEs.

By 45°C, the A_4H8-H1' crosspeak has disappeared (boxed region in figure 5c). This data, along with results from the imino proton spectra, indicates that although the terminal AT base pairs remain in the Hoogsteen conformation throughout the temperature range studied, the internal AT base pairs are exchanging between a Hoogsteen base paired and an unpaired (or Watson-Crick base paired) state at the higher temperatures. Thus, internal Hoogsteen AT base pairs appear to be relatively unstable in this echinomycin-DNA complex, and may not be the relevant conformation *in vivo*.

SUMMARY

We have presented evidence for two different biologically relevant DNA conformations which contain Hoogsteen base pairs. In the triplex, Hoogsteen base pairs are formed when the $d(TC)_4$ strand binds in the major groove of a DNA duplex formed from $d(GA)_4$ and $d(TC)_4$. This binding stabilizes the DNA. The TA Hoogsteen base pairs appear to be more stable than the C^+G Hoogsteen base pairs, probably as a result of the requirement for base protonation to form Hoogsteen C^+G base pairs. In the echinomycin-DNA complex, Hoogsteen base pairs are formed adjacent to the echinomycin binding sites. The internal Hoogsteen base pairs, which are constrained by the helix, appear to be kinetically less stable than terminal AT Hoogsteen base pairs. Thus, in the absence of torsional helical constraints, Hoogsteen base pairing is a viable alternative to Watson-Crick base pairing in DNA.

ACKNOWLEDGMENTS

The authors thank Drs. J.H. van Boom and G.A. van der Marel for providing some of the DNA oligonucleotides used in these studies and Dr. A.H.-J. Wang for many helpful discussions. Echinomycin was a gift from the National Cancer Institute.

REFERENCES

1. Hoogsteen K (1959). The structure of crystals containing a hydrogen-bonded complex of 1-methylthymine and 9-methyladenine. Acta Cryst 12:822.
2. Wells RD, Collier DA, Hanvey JC, Shimizu M, Wohlrab F (1988). The chemistry and biology of unusual DNA structures adopted by oligopurine oligopurine sequences. FASEB J 2:2939.
3. Cantor CR, Efstratiadis A (1984). Possible structures of homopurine·homopyrimidine S1-hypersensitive sites. Nucl Acids Res 12:8059.
4. Pulleyblank DE, Haniford DB, Morgan AR (1985). A structural basis for S1 nuclease sensitivity of double-stranded DNA. Cell 42:271.
5. Mirkin SM, Lyamichev VI, Drushlyak KN, Dobrynin VN, Filippov SA, Frank-Kamenetskii MD (1987). DNA H form requires a homopurine-homopyrimidine mirror repeat. Nature 330:495.
6. Moser HE, Dervan PB (1987). Sequence-specific cleavage of double helical DNA by triple helix formation. Science 238:645.
7. Rajagopal P, Feigon J (1989). Triple strand formation in the homopurine:homopyrimidine DNA oligonucleotides $d(GA)_4$ and $d(TC)_4$. Nature: in press.

8. Rajagopal P, Feigon J (1989). NMR Studies of Triple Strand Formation From the Homopurine: Homopyrimidine Deoxyribonucleotides d(GA)$_4$ and d(TC)$_4$. Biochemistry: accepted.
9. Katagiri K, Yoshida T and K (1974). In Corcoran JW, Hahn FE (eds): "Antibiotics III: Mechanism of Action of Antimicrobial and Antitumour Agents", Berlin: Springer, p. 234.
10. Waring MJ (1979). In Hahn FE (ed): "Antibiotics V/Part 2: "Mechanism of Action of Antieukaryotic and Antiviral Compounds", Berlin: Springer, p. 173.
11. Low CML, Drew HR, Waring, MJ (1984) Sequence-specific binding of echinomycin to DNA: evidence for conformational changes affecting flanking sequences. Nucl Acids Res 12:4865.
12. Van Dyke MM, Dervan PB (1984). Echinomycin binding sites on DNA. Science 225:1122.
13. Ughetto G, Wang, AH-J, Quigley GJ, van der Marel GA, van Boom JH, Rich A (1985). A comparison of the structure of echinomycin and triostin A complexed to a DNA fragment. Nucl Acids Res 13:2305.
14. Wang AH-J, Ughetto G, Quigley GJ, Hakoshima T, van der Marel GA, van Boom JH, Rich A (1986). The molecular structure of a DNA-Triostin A complex. Science 225:1115.
15. Quigley GJ, Ughetto G, van der Marel GA, van Boom JH, Wang AH-J, Rich A (1986). Non-Watson-Crick G·C and A·T base pairs in a DNA-antibiotic complex. Science 232:1255.
16. Gilbert DE, van der Marel, GA, van Boom JH, Feigon, J (1989). Unstable Hoogsteen basepairs adjacent to echinomycin binding sites within a DNA duplex. PNAS 86:3006.
17. Gilbert DE, Feigon J (1989). Manuscript in preparation.
18. Feigon J, Wright JM, Leupin W, Denny WA, Kearns DR (1984)). Use of two-dimensional NMR in the study of a double-stranded DNA decamer. J Am Chem Soc 104:5540.

Frontiers of NMR in Molecular Biology, pages 259-270

Isotope Directed 2-Dimensional NMR Experiments: Applications to Studies of DNA Conformation

David H. Live

Department of Chemistry, Emory University
Atlanta, Georgia 30322

ABSTRACT Although it has been generally impractical to incorporate isotopic labels and implement isotope directed experiments in deoxyoligonucleotides, the sensitivity advantages of ^{1}H indirect detection techniques make it possible to study ^{31}P, ^{15}N and ^{13}C at natural abundance in such systems. Here we report the use of NMR parameters from these nuclei for conformational analysis of DNA.

INTRODUCTION

Within a short period of their introduction, isotope directed methods have proven very valuable if not indispensable in the study of biopolymers.(1,2) These methods have found their greatest application in the study of proteins and polypeptides. Although some results have been reported at natural abundance (3,4), most applications have been facilitated by comparatively easy methods for incorporation of isotopic labels biosynthetically (5) or via peptide synthesis techniques. Techniques from molecular biology of cloning the gene for the protein of interest and expressing it in wild type or auxotrophic organisms grown on labeled media have been used for biosynthesis. Isotopically labeled amino acids can also be readily converted to the protected forms needed for chemical synthesis of polypeptides. ^{2}H, ^{13}C, and ^{15}N incorporated in these ways have generally served as aids in spectral editing for 1D, 2D and 3D NMR experiments as illustrated in some of the other contributions to this volume. In particular, in 2D and 3D experiments the isotopes often function to provide an additional mode for achieving spectral dispersion. From these experiments, the specific ^{13}C and ^{15}N shifts have been assigned in several proteins, however, the quantitative interpretation of such shifts in structural terms has been elusive.(5)

In contrast, the application of isotope directed techniques to the study of oligonucleotides, discussed here, is less extensive in large part because the advantages afforded by isotopic labeling have not been readily available. Isotopic enrichment is far more difficult in these systems. Such studies can still be carried out effectively for oligonucleotides at natural abundance (6-13) with the sensitivity advantages of ^{1}H indirect detection, and the results show promise of elucidating conformational properties, as is described below. The three heteronuclei of interest in these molecules are ^{31}P, ^{15}N, and ^{13}C.

RESULTS AND DISCUSSION

^{31}P Studies

^{31}P is distinct among these isotopes in that it naturally occurs at 100% abundance. This factor along with the relatively high sensitivity of this nucleus expand the number of experimental options available. It was already realized some time ago that the ^{1}H - ^{31}P HMQC experiment could be applied to oligonucleotides to aid in making sequential assignments. (11) Such an experiment allows one to walk along the sugar phosphate backbone through connectivities indicated in the 2D plot, since there is coupling between C3' protons on one sugar and C5' protons on the next sugar to the intermediate phosphate. The ^{31}P resonance of the phosphate is assigned at the same time. This provides an alternative means of sequential assignment in oligonucleotides. Such an approach can be important in cases where irregular structures exist and the normal sequential base to base NOE's used in assignment may be disrupted. Since the ^{31}P shifts of the phosphate groups can themselves be indicators of unusual conformations, this helps in pinpointing such locations. More recently a ^{1}H detected ^{1}H -^{31}P phase sensitive COSY experiment has been developed which can provide the same information, and is readily used in the phase sensitive mode. (12) Although this experiment is in principle less sensitive than the HMQC experiment,since one observes the lower ^{31}P magnetization transferred to the protons, it typically provides better resolution in the ^{1}H dimension, a characteristic that is valuable when making correlations to crowded H3', H4' and H5' and H5" regions. A further extension of this experiment makes it possible to extract the vicinal coupling between the ^{31}P atom and the H3', helping to define an angle in the sugar phosphate backbone. (13) To accomplish this, a selective 180° pulse is applied to the H3' protons in the middle of the evolution period. Such a selective inversion refocuses the effect of the spin coupling interaction during this period with the result that the splitting of the anti-phase doublet in the phosphorous dimension of the cross peak is the vicinal coupling. This works because the H3' protons are separated from the other protons to which they have homonuclear coupling. As has also been pointed out, the use of a selective 180° pulse to decouple the H3' protons from their coupled proton partners can be applied to a ^{1}H detected 2DJ experiment.(13) The latter experiment offers better sensitivity than the COSY, but is subject to greater interference from the residual HOD peak and the overlap of H3' resonances. On the positive side, the problem of correcting the coupling value obtained from splitting of antiphase doublet pair, found in the COSY spectrum, for the effects of linewidth is not necessary with the 2DJ. The combined information from unusual shifts and coupling constants immediately indicate sites of unusual conformation. Since such sites are often lacking in the usual NOE contacts,(14) the alternative information such as vicinal couplings are an aid in counteracting the deficit in conformational constraints needed to solve the structure.

Both the selective COSY experiment and the 2DJ experiment have been utilized in analyzing the backbone vicinal coupling in a regular DNA duplex structure. An extension of this is to apply these experiments in systems of irregular nature, such as possible hairpin structures, where changes in these angles would be expected. An example of this has been in an ongoing study of a Pt drug DNA complex in collaboration with Luigi Marzilli's laboratory at Emory.(15) The cis Pt complex from reacting d(TATGGGTACCCATA) with Pt(ethylenediamine)Cl_2 was examined as a model for the interaction of this important anti-cancer drug with DNA. A variety of physical measurements (UV and CD spectroscopy and electrophoresis) as well as NMR indicate that the molecule does not form a duplex. The ^{31}P signals were assigned using ^{17}O labeling and combined ^{1}H homonuclear

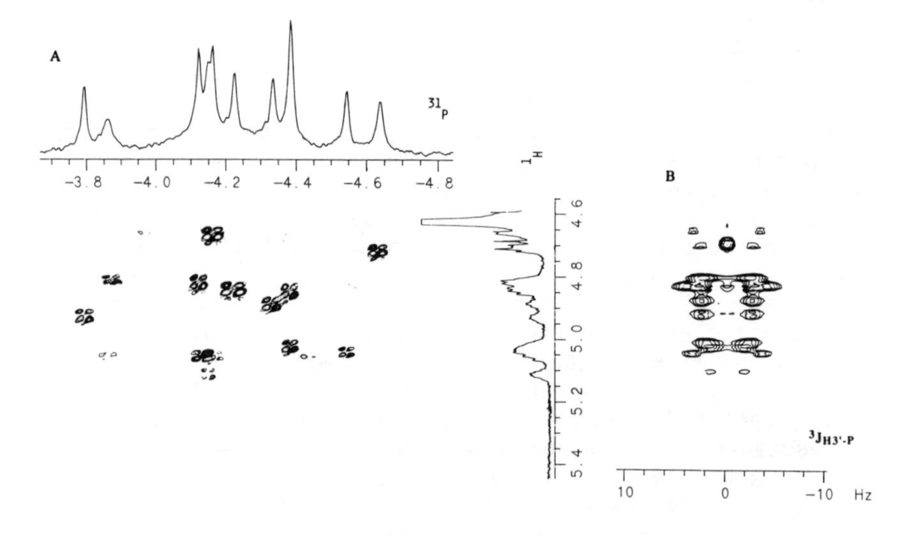

Figure 1. Sections of the contour plots of A) ^{1}H detected ^{1}H-^{31}P selective reverse chemical shift correlation experiment (13), and B) ^{1}H detected ^{1}H-^{31}P heteronuclear 2DJ experiment (13) of d(TATGGTACCCATA)-Pt(en) (3 mM in D_2O, pD 7.4, 40°C). 1D ^{1}H and ^{31}P are presented along their respective axes. For A, 128 increments of 128 scans each were collected for a sweep width in the ^{31}P dimension of 300 Hz. The magnitude of the selective pulse was 125 Hz centered at 4.8 ppm. Alternate scans were stored in separate regions of memory and processed to yield hypercomplex phase sensitive presentation. (13) The multiplet splitting in the ^{31}P dimension is the H3'-P vicinal coupling. The spectral width of the ^{31}P dimension was reduced to enhance digital resolution. This gave rise to fold over of the signal with actual ^{31}P shift of -2.65 ppm, to -4.25 ppm ^{31}P and 5.05 ppm ^{1}H. The signal normally at -3.00 ppm is not seen in this experiment since its coupling is very small. For B, 64 increments of 64 scans were collected for the heteronuclear 2DJ experiment with a spectral width in f_1 of 50 Hz. The magnitude of the selective pulse, centered at 4.75 ppm was 119 Hz. Data were collected on a GN-500 spectrometer. Reproduced from ref. 15 with permission.

2-D assignments with ^{1}H detected heteronuclear 2D experiments. Selective COSY and 2DJ experiments (13) were carried out (figure 1) to determine the H3' to ^{31}P couplings in the sugar phosphate backbone. Three of the proton - phosphorous couplings were found to have unusual values relative to what was observed in a normal duplex. These are those to G5 (less than 1Hz), A8 (6.8Hz), and C9 (8.0Hz). This corresponds to an opening of the torsion angle at G_5, and a closing of it at the other two sites. Two of the associated ^{31}P resonances were shifted downfield from the normal positions, G5 at -3.00 ppm and A8 at -2.65 ppm (relative to trimethylphosphate). The changes in the bond angles add further support to the contention that the Pt binding and distortion in structure occurs in this part of the molecule.

^{13}C Studies

^{13}C natural abundance studies of oligonucleotides present more formidable problems than those for ^{31}P, since sensitivity is down by approximately two orders of magnitude. Nonetheless, ^{13}C spectra of DNA fragments have been obtained. Spectra of several duplexes have been reported recently, mostly using proton detected heteronuclear multiquantum (HMQC) methods. (6-10, 16) The molecules examined have generally been assumed to adopt B DNA structure. It has already been noted that the ^{1}H - ^{13}C 2-D experiments should aid in making ^{1}H assignments since there is much greater dispersion between the shifts for different types of ^{13}C than for ^{1}H's.(6) Particularly in regions where classes of protons overlap, for instance where the sugar H3', H4', H5' and H5" protons resonate, the large difference in the ^{13}C shifts, over 20 ppm, may help clarify ambiguities in proton assignments.

The effect of duplex melting on the shifts has also been investigated, (16) and in one instance the transition from B to Z DNA has been examined.(9) Certainly the greater dispersion of ^{13}C shifts and their relative sensitivity to conformational effects suggests that this parameter could be used at least qualitatively to identify unusual structural features in DNA oligomers. To assess this we have examined the shifts of the C1' sugar carbons of several oligomers thought to adopt hairpin structures. One would expect to see significant effects at this position since this is the point of attachment of the base, and should be sensitive to changes in both sugar pucker and base orientation. On melting and in the B to Z transition, shifts of several ppm downfield are reported for the anomeric carbon.

For the four DNA sequences assumed to be in B-DNA

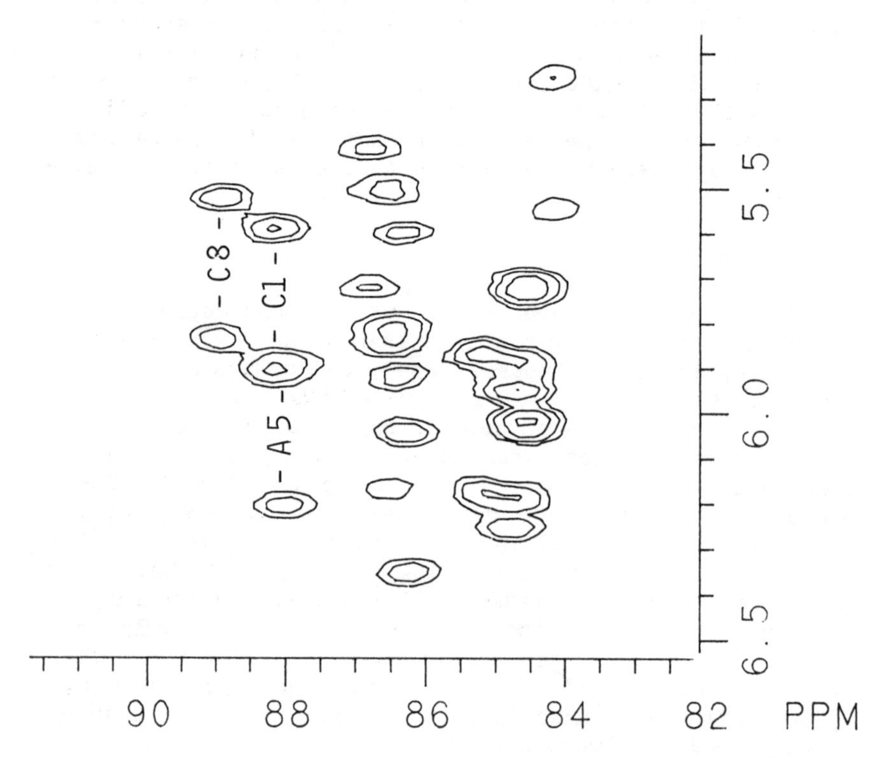

Figure 2. Section of the 2-D contour plot of a ^{1}H detected ^{1}H-^{13}C spectrum of d(CGCGAACCATCGCG) showing the anomeric carbons and protons from the sugars. Sample contained 400 OD units of DNA in 0.4 ml D_2O at pD 6.4. The HMQC experiment with refocusing during the evolution period but no decoupling was used. (20) This gives rise to two cross peaks for each carbon- proton pair, one from each proton satellite. Data were collected in 128 increments of 512 scans each with a spectral width in the f_1 dimension of 8064 Hz. Assignments of some of the pairs of cross peaks are indicated. Shifts are relative to TSP in both dimensions. Spectra were obtained on a GN-500 spectrometer.

conformation where assigned ^{13}C shifts have been reported, it has been found that the C1' shifts fall between 84 and 87 ppm relative to TSP. The only deviation from this range for B-DNA samples is when the C1' is on a pyrimidine base at position 1. (8,16) Poly d(Gm^{5}C), one of the oligos referred to above, has also been converted to the Z form by addition of salt and elevation of temperature.(9) On transformation to the Z form the C1's shift approximately 2.5 ppm downfield to shifts greater than 87 ppm. For the G residue in this sequence the base is reorienting from an anti to a syn conformation relative to the sugar. The only available data on the effect of melting on ^{13}C sugar carbons is for the sequence [d(CG)$_3$]. (16) Under conditions where the original duplex is apparently melting into single strands, the internal anomeric carbons shift downfield by approximately 1.5 ppm. Consequently it appears that changes in DNA structure induce shifts at C1' ^{13}C, and we have investigated the C1' shifts of several DNA's that appear to adopt hairpin structures, including the Pt(en)DNA complex discussed above, to extend this correlation. In each of these cases we have found several resonances occurring downfield of 87 ppm, the normal limit of the range for B-DNA C1's. In the spectrum (figure 2) for the sequence d(CGCGAACCATCGCG) where a loop of the central ACCA sequence is postulated, the C1's of A5 and C8 are found at 88.22 and 89.14 ppm from TSP respectively. The other C1' resonance in this region is from the residue C1 at 88.34 ppm. For one of the residues in the Pt-DNA complex, the shift of the C1' of a G residue, known to be syn from base to sugar NOE's, is well downfield at about 90 ppm along with several others below 87 ppm. Although at this time the factors contributing to ^{13}C shifts can not be quantitated, it seems clear from these data that the unusual shifts correlate with unusual sugar-phosphate backbone conformation relative to B-DNA.

^{15}N Studies

Proton indirect detection of ^{15}N is the most challenging of the experiments discussed here since ^{15}N has the lowest natural abundance. These studies also must be carried out in water since the protons of interest for detection are exchangable. It was pointed out some time ago that the HMQC experiment could be carried out with a single proton pulse, and a Redfield 2-1-4-1-2 pulse that minimizes excitation of water can be employed. (1) Studying the ^{15}N

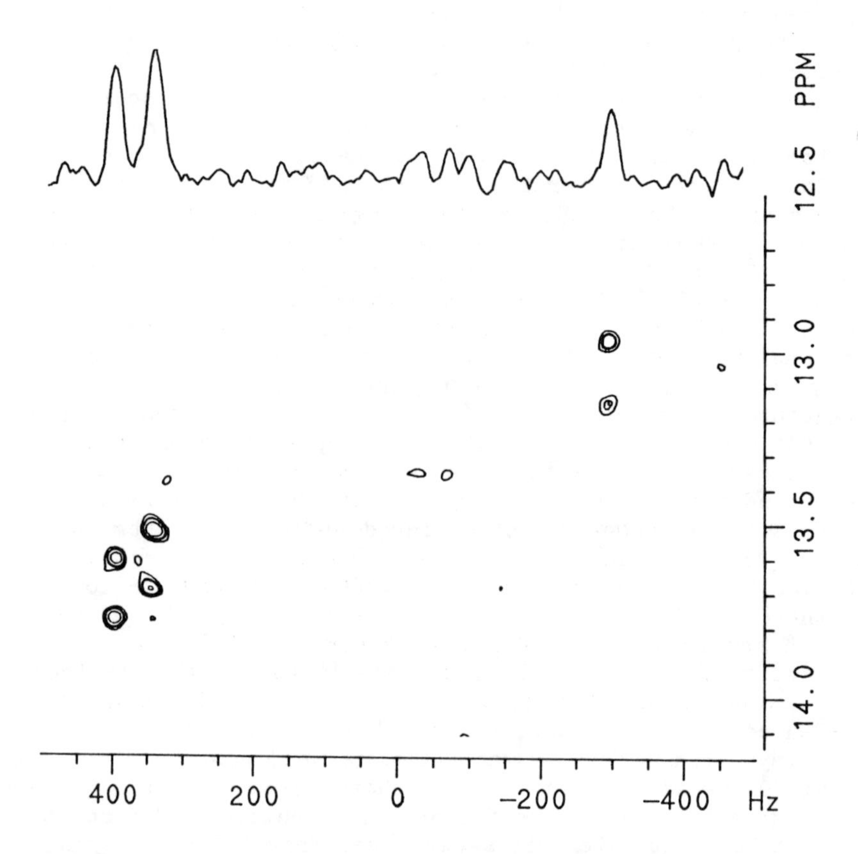

Figure 3. 2-D contour plot from the imino region of a ^{1}H detected ^{1}H - ^{15}N HMQC experiment on d(GGAATTCC)$_2$ in H_2O at neutral pH. The concentration of strands was 8 mM in 0.4 ml. 30 increments of 2048 scans were taken in t_1 for a sweep width of 1000 Hz. The HMQC experiment was used with ^{1}H excitation by a single Redfield pulse (20). Spectra were obtained on a modified NT-300 spectromenter. Two cross peaks are observed for each ^{15}N, one for each ^{1}H satellite. The vertical axis shows the ^{1}H shift, while the horizontal axis, which contains the ^{15}N information is in units of the multiquantum frequency, the sum of ^{1}H and ^{15}N offsets. From this and the known proton position the ^{15}N shift is readily obtained. A projection in the multiquantum direction essentially shows the sensitivity of ^{15}N detection.

spectrum of imino protons in oligonucleotides should be valuable since these nuclei are in a key position to report on the interstrand hydrogen bonding. We have employed this approach in studying the imino nitrogens of the duplex d(GCTATACC)$_2$, and a spectrum is shown in figure 3. This duplex had already been studied extensively by proton NMR (17) so the nitrogen resonances could be assigned. Shifts of the imino nitrogens on the G and C residues are similar to those previously reported for large DNA pieces, biosynthetically enriched with ^{15}N, of 151 and 154 ppm (relative to NH_3) for G and C respectively in solution (18) and 148 and 160 ppm respectively in the solid.(19) Comparison with individual nucleotides in solution is a bit risky since these shifts should be very sensitive to hydrogen bonding, and the arrangement in the duplex is likely to be different than hydrogen bonding to a solvent.

The duplex was studied with and without the addition of the drug netropsin. Three of the four distinct imino nitrogens could be observed in each case. The terminal imino proton was broad, presumably from rapid exchange due to fraying at the end of the fragment, and this impeded the observation of its attached nitrogen. The N1 of G2 has a shift of 147.2 in the normal duplex and 147.47 in the netropsin complex. The T6 N3 and T5 N3 have shifts of 155.93 and 155.50 ppm (relative to NH_3) in the duplex, respectively, but in the drug complex the shifts are 160.03 and 158.66 ppm respectively. The larger nitrogen shift changes occur near the center of the DNA fragment where the netropsin binds. (21) It is thought that the netropsin binds in the minor groove, displacing waters hydrogen bonded to the bases. The removal of hydrogen bonding to the water must be replaced either by hydrogen bonding to netropsin, or strengthening of the interstrand hydrogen bonding since the downfield shift indicates increased hydrogen bonding.

CONCLUSIONS

Although the use of parameters from heteronuclear experiments on oligonucleotides is in its early stages compared to application of homonuclear proton 2-D methods, it is clear that such experiments, even at natural abundance, can assist in the conformational analysis of DNA fragments and their interactions with other molecules. ^{31}P and ^{13}C related experiments can address variations in the sugar phosphate backbone, while ^{15}N experiments can elucidate the specifics of modulation of the hydrogen bonding interactions.

ACKNOWLEDGMENTS

This work was supported by research and instrumentation grants from NIH and NSF.

REFERENCES

1. Griffey, RH, Redfield, AG (1987). Proton-Detected Heteronuclear Edited and Correlated Nuclear Magnetic Resonance and Nuclear Overhauser Effect in Solution. Q Rev Biophys 19:51-82.

2. Bax, A (1989). Two-Dimenisonal NMR and Protein Structure. Ann Rev Biochem 58:223-56.

3. Ortiz-Polo G, Krishnamoorthi R, Markley JL, Live DH, Davis DG, Cowburn D (1986). Natural Abundance ^{15}N Spectroscopy of Turkey Ovomucoid Third Domain: Assignment of Peptide ^{15}N Resonances of Residues at the Reactive Site via Proton-Detected Multiple Quantum Coherence.J Magn Reson 68:303-310. and Glushka J, Cowburn D (1987). Assignment of ^{15}N Shifts in BPTI. J Amer Chem Soc 109:7879.

4. Bruhwiler D, Wagner G (1986). Toward the Complete Assignment of the Carbon Nuclear Magnetic Resonance Spectrum of Basic Pancreatic Tripsin Inhibitor. Biochemistry 25:5839-5843.

5. Muchmore DC, McIntosh LP, Russell CB, Anderson DE, Dahlquist FW (1989). Expression and ^{15}N Labeling of Proteins for ^{1}H and ^{15}N NMR. "Methods in Enzymology" 177 New York: Academic Press 44-73.

6. Leupin W, Wagner G, Denny WA, Wuthrich K (1987). Assignment of the ^{13}C Nuclear Magnetic Resonance of a Short DNA-Duplex with ^{1}H-detected Two-Dimensional Heteronuclear Correlation Spectroscopy. Nucl Acids Res 15:267-275.

7. Ashcroft J, LaPlante SR, Borer PN, Cowburn D (1989). Sequence Specific ^{13}C NMR Assignment of Non-protonated Carbons in $[d(TAGCGCTA)]_2$. J Amer Chem Soc 111:363.

8. LaPlante SR, Ashcroft J, Cowburn D, Levy GC, Borer PN (1988). ^{13}C NMR Assignments of the Protonated Carbons of $[d(TAGCGCTA)]_2$ by Two-Dimensional Proton-Detected Heteronuclear Correlation. J Biomol Struc and Dynam 5:1089-1099.

9. Sklenar V, Bax A (1987). Assignment of Z DNA NMR of Poly $d(Gm^5C)$ by Two-Dimensional Multinuclear Spectorscopy. J Amer Chem Soc 109:2221-2222.

10. LaPlante SR, Boudreau, Zanatta N, Levy GC, Borer PN (1988). ^{13}C-NMR of the Bases of Three DNA Oligonucleotide Duplexes: Assignment Methods and Structural Features. Biochem 27:7902-7909.

11. Byrd RA, Summers MF, Zon G, Fouts CS, Marzilli LG (1986). A New Approach for Assigning ^{31}P NMR Signals and Correlating Adjacent Deoxyribose Moieties via ^{1}H Detected Multiple-Quantum NMR. Application to the Adduct of d(TGGT) with the Anticancer Agent, (ethylenediamine) dichloroplatinum. J Amer Chem Soc 108:504-505.

12. Sklenar V, Miyashiro H, Zon G, Miles HT, Bax A (1986). Assignment of the Phosphorous-31 and Proton Resonances of Oligonucleotides by Two-Dimensional NMR. FEBS Lett 208:94-98.

13. Sklenar V, Bax A (1987). Measurement of ^{1}H-^{31}P NMR coupling Constants in Double-Stranded DNA Fragments. J Amer Chem Soc 109:7525-7526.

14. Neuhaus D, Williamson M (1989)."The Nuclear Overhauser Effect in Structural and Conformational Analysis" New York: VCH Publishers, Chap. 12.

15. Marzilli LG, Kline TP, Live D, Zon G (1989). ^{31}P NMR Spectroscopic Investigations of Pt-anticancer Drug Binding to Oligonucleotide Models of DNA, the Probable Molecular Target. In Tullius T (ed) "DNA-Metal Chemistry" ACS Symposium Series 402 Washington, ACS p. 119.

16. Borer PN, Zanatta N, Holak TA, Levy GC, van Boom JH, Wang AH-J (1984). Conformation and Dynamics of Short DNA Duplexes: $(dC\text{-}DG)_3$ and $(dC\text{-}dG)_4$. J Biomol Struc and Dynam 1:1373-1386.

17. Patel DJ, Shapiro L, Hare D (1986). Sequence-dependent Conformation of DNA COmplexes. The AATT Segment of the d(GGAATTCC) Duplex in Aqueous Solution. J Biol Chem 261:1223-1229.

18. James TL, James JL, Lapidot A (1981). Structural and Dynamic Information about Double Stranded DNA from Nitrogen-15 NMR Spectroscopy. J Amer Chem Soc 103:6748-6750.

19. Cross TA, DiVerdi JA, Opella SJ (1982). Strategy for Nitrogen NMR of Biopolymers. J Amer Chem Soc 104:1759-1761.

20. Bax A, Griffey RH, and Hawkins BL (1983). Correlation of Proton and Nitrogen-15 Chemical Shifts by Multiple Quantum NMR. J Magn Reson 55:301-315.

21. Kopka ML, Yoon C, Goodsell D, Pjura P, Dickerson RE (1985). The Binding of Netropsin to Double-Helical B-DNA of Sequence $CGCGAATT^{Br}CGCG$. Single Crystal X-ray Structural Analysis. Proc Natl Acad Scis USA 82:1376-1380.

Frontiers of NMR in Molecular
Biology, pages 271-274

DISCUSSION SUMMARY: DNA/RNA AND LIGAND INTERACTIONS

Thomas L. James

Departments of Pharmaceutical Chemistry and Radiology
University of California, San Francisco, CA 94143

Sequence-dependent structural variations in nucleic acids are important for the protein-nucleic acid and RNA-DNA interactions essential for life. Recent progress in synthesis techniques have enabled synthetic DNA and RNA fragments to be prepared. These oligomers have been subsequently utilized in structural studies using x-ray crystallography, NMR or other spectroscopic methods. Furthermore, they have been used for studies of protein binding and drug binding. Concomitant developments in molecular biology have enabled larger DNA molecules with specified sequence to be prepared and utilized for complementary studies.

With advances in both the ability to prepare desired nucleic acids and in modalities suitable for probing their structure, substantial polymorphism in DNA structure has been found. Examples of DNA structural variations are shown in Figure 1. Not all of these structural variants have been demonstrated yet *in vivo*. The short range interactions which influence measurable NMR parameters are quite sensitive to the local molecular environment, but NMR is less suitable for discerning the larger structural motifs. This particular session of the symposium deals with local structural features in nucleic acids and in complexes with proteins and drugs. Abstracts of the presentations of Drs. R. Kaptein, A. Pardi, and D. Wemmer may be found in the *Journal of Cellular Biochemistry*, Supplement 13A, 1989.

Dr. Robert Kaptein described work with *lac* repressor, lex A repressor, and Arc repressor. Methodology entailed 2D NMR and homo-

a.

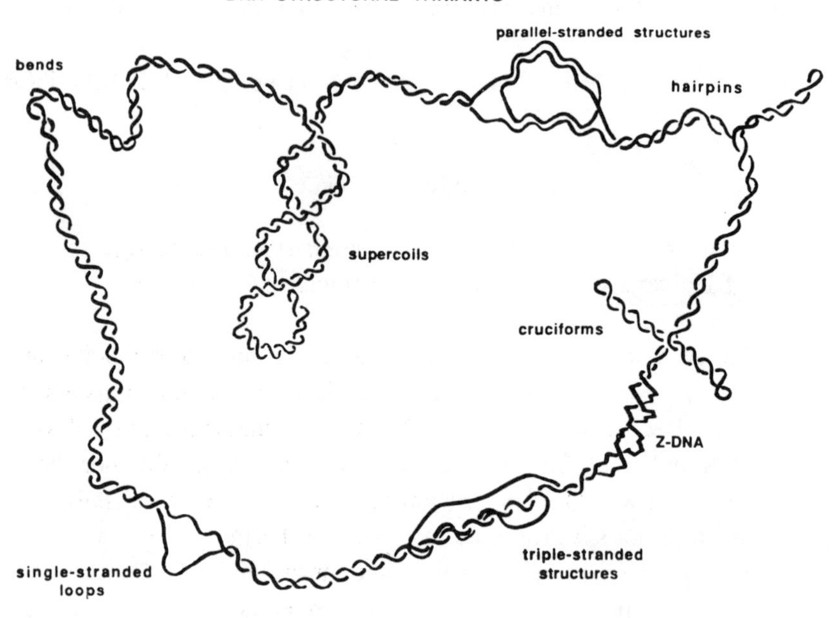

b.

FIGURE 1. DNA structural variations. Large-scale polymorphic structural features (a) are reflected in torsion angle variations in the individual nucleotide residues (b).

nuclear 3D NMR experiments. Most work has been carried out with *lac* repressor headpiece and with the 22 base-pair *lac* operator or with 11 or 14 base-pair half operators. Following resonance assignments, structural features could be deciphered. Repressor-operator complex formation yielded minimal conformational changes in either protein or DNA. Intranucleotide NOE's delineated amino acid-base contacts and permitted computer modeling and depiction of the structure of the complex.

Dr. Arthur Pardi discussed NMR studies on RNA folding domains. Specifically, they have synthesized some "hammerhead" RNA oligomers which have the capacity for self-cleavage. These particular molecules contain three helices, require a divalent cation, and require a 2′-hydroxyl at the cleavage site. A non-cleaving oligonucleotide hybrid suitable for study was made in which ¾ of the molecule is RNA, comprising the catalytic moiety, and ¼ of the molecule is DNA, in place of the substrate moiety. Early studies of the secondary and tertiary structure were presented.

Dr. Dinshaw Patel described studies of the complex formed between the anti-cancer agent chromomycin and $[d\text{-}(AAGGCCTT)]_2$. NMR evidence for binding of the drug dimer to the minor groove, which was altered to be wide and shallow as in A-form DNA, was presented. Preliminary studies of triple-stranded DNA helices were also presented. The oligonucleotide $[d\text{-}(CTCCCTCTCCC)]_2 \cdot d\text{-}(GAGGGAGAGGG)$ at low pH (5.0) was found to have imino proton resonances at frequencies normally characteristic of Watson-Crick base pairing. But evidence of Hoogsteen base pairing between the purine strand and the additional pyrimidine strand was manifest by the appearance of additional imino proton resonances at 8.5 - 10 ppm.

Dr. David Wemmer presented research results on the interaction of the anti-cancer drug distamycin-A with DNA oligomers. NMR evidence was given indicating that distamycin binds in the minor groove and, with some sequences, notably $[d\text{-}(CGCAAATTGCG)]_2$, the drug can rapidly slide back and forth in the minor groove to cover five AT base pairs rather than the four which would correspond to the size of distamycin. Although the drug prefers AT sequences, it was also demonstrated that interactions of the drug with GC-containing sites could also take place when the AT-containing sites were saturated.

Frontiers of NMR in Molecular Biology, pages 275-276

DISCUSSION SUMMARY: LARGER SYSTEMS/ORDERED SYSTEMS

S. J. Opella

Department of Chemistry
University of Pennsylvania
Philadelphia, Pennsylvania 19104

The success of multi-dimensional NMR studies of biopolymers in solution was demonstrated in presentations throughout this Symposium. Although the presentations were concerned with a wide variety of protein and nucleic acids, all of the systems had one common feature and that was that the molecule of interest reorients rapidly in solution. This session, in contrast, was concerned with systems that do not have convenient reorientation properties for studies in solution, because the molecules are in the crystalline or hydrated solid state, part of supramolecular structures in solution, or are simply too large. These systems are of considerable interest because many biological functions are expressed with large and immobile molecules and their complexes, for example membrane bound proteins or protein-nucleic acid complexes. It is essential to develop NMR approaches for their investigation, since these systems tend to be difficult to crystallize in forms suitable for x-ray diffraction analysis and they are not amenable to conventional NMR methods for solutions.

There was a unifying theme to this session, even though the systems discussed included peptides, proteins, DNA, and lipids and the spectroscopic techniques used to investigate these systems included those generally associated with single crystals, powder, liquid crystals, in addition to those for solutions. The theme was that innovative isotopic labelling schemes and the full range of NMR experiments are needed to study biological systems in systems where the molecule of interest does not reorient rapidly in solution.

S. J. Opella presented the results of studies on the membrane bound forms of the procoat and coat proteins of the filamentous bacteriophages. Solid-phase peptide synthesis was used to prepare samples of the procoat protein with ^{15}N labels incorporated in specific sites. This enabled direct comparisons of the structure and dynamics of the procoat and coat proteins, with and without the N-terminal leader sequence. The dynamics of backbone sites were described with motionally averaged ^{15}N powder pattern lineshapes for the proteins in phospholipid bilayers and with the heteronuclear $^{1}H/^{15}N$ NOE for the proteins in micelles in solution. Even though these proteins reorient relatively slowly in micelles, uniform labelling with ^{2}H and ^{15}N enabled two-dimensional NMR experiments to be useful for structural studies.

J. Prestegard described a new approach for investigating membrane associated glycolipids. The surface headgroups of these lipids have important biological roles for receptors, viruses, toxins, and cell differentiation. In this oriented systems, structural informatiion is available from the angular dependence of quadrupolar and dipolar couplings available for analysis from ^{2}H labelled sites. Significantly, it was possible to separate motional from structural effects. The approach has sufficient sensitivity and resolution to clearly differentiate bound and unbound conformations of the glycolipid headgroup in the presence and absence of proteins, such as the b subunit if Ricin.

P. Tsang described solution NMR studies of a relatively large system, the complex between monoclonal anti-peptide Fab' molecules and their complementary peptides. These studies were made possible by ^{15}N labelling of the peptides so that their ^{1}H resonances could be monitored without interference from those of the much larger protein. This is an example of isotope-edited NMR spectroscopy. They were able to detect significant difference in the NMR spectra from the various labelled peptide sites that could be correlated with what is already known about the epitope region of this peptide.

G. Drobny described solid-state NMR studies of DNA. These studies emphasized the analysis of the dynamics of the polymer as a function of hydration. The analysis used both lineshapes and relaxation parameters. The interpretation of the data was discussed in terms of local and long range motions in DNA and how they are influenced by the conformational state of the molecule. A thorough analysis is possible because of the availability of data from multiple deuterium labelled sites in the bases, sugars, as well as the phosphodiester backbone.

G. Marshal presented an application of a novel solid-state NMR experiment to determining distances in a synthetic peptide. This technique, called REDOR, enables the measurement of weak heteronuclear dipolar couplings in polycrystalline samples. By throughtful selection of ^{13}C and ^{15}N labelled sites and exploiting the high precision of this experiment, they were able to distinguish between α-helix and 3_{10}-helix in the peptide.

Index